Springer Laboratory

Springer Laboratory Manuals in Polymer Science

Pasch, Trathnigg: HPLC of Polymers
ISBN: 3-540-61689-6 (hardcover)
ISBN: 3-540-65551-4 (softcover)

Mori, Barth: Size Exclusion Chromatography
ISBN: 3-540-65635-9

Pasch, Schrepp: MALDI-TOF Mass Spectrometry of Synthetic Polymers
ISBN: 3-540-44259-6

Kulicke, Clasen: Viscosimetry of Polymers and Polyelectrolytes
ISBN: 3-540-40760-X

Hatada, Kitayama: NMR Spectroscopy of Polymers
ISBN: 3-540-40220-9

Brummer, R.: Rheology Essentials of Cosmetics and Food Emulsions
ISBN: 3-540-25553-2

Mächtle, W., Börger, L.: Analytical Ultracentrifugation of Polymers and Nanoparticles
ISBN: 3-540-23432-2

Heinze, T., Liebert, T., Koschella, A.: Esterification of Polysaccharides
ISBN: 3-540-32103-9

Koetz, J., Kosmella, S.: Polyelectrolytes and Nanoparticles
ISBN: 3-540-46381-X

Striebeck, N.: X-Ray Scattering of Soft Matter
ISBN: 3-540-69855-5

Schärtl, W.: Light Scattering from Polymer Solutions and Nanoparticle Dispersions
ISBN: 3-540-71950-2

Khulbe, K.C., Feng, C.Y., Matsuura, T.: Synthetic Polymeric Membranes
ISBN: 3-540-73993-7

Kailash C. Khulbe · C.Y. Feng · Takeshi Matsuura

Synthetic Polymeric Membranes

Characterization by Atomic Force Microscopy

Springer

Dr. Kailash C. Khulbe
C.Y. Feng
Dr. Takeshi Matsuura

Industrial Membrane Research Laboratory
Chemical Engineering Department
University of Ottawa
Louis Pasteur St. 161
Ottawa, K1N 6N5
Canada

khulbe@eng.uottawa.ca
cyfeng@eng.uottawa.ca
matsuura@eng.uottawa.ca

ISBN 978-3-540-73993-7 e-ISBN 978-3-540-73994-4

DOI 10.1007/978-3-540-73994-4

Library of Congress Control Number: 2007931849

Cover design: WMXDesign GmbH, Heidelberg, Germany
Typesetting and production: LE-TEX Jelonek, Schmidt & Vöckler GbR, Leipzig, Germany

Printed on acid-free paper

9 8 7 6 5 4 3 2 1

springer.com

Springer Laboratory Manuals in Polymer Science

Preface

This book concentrates on atomic force microscopy (AFM), a method recently developed to study the surfaces of synthetic polymeric membranes. AFM is becoming a very important tool for the characterization of synthetic polymeric membranes. The development of membranes of improved performance depends on the exact knowledge of the morphology of a thin selective layer that exists at the surface of the membrane. The control of the morphology of the selective layer is crucial for the design of synthetic polymeric membranes. With a relatively short history of only twenty-five years, AFM has firmly established its position as a method to characterize the membrane surface.

Each chapter of this book includes information on basic principles, commercial applications, current research, and guidelines for future research. Each chapter is summarized at the end and contains a comprehensive list of references.

The introductory chapter gives a brief overview of synthetic polymeric membranes and their applications both in industrial processes and in biomedical fields. It also gives an overview of studies on membrane surface morphology by various methods.

Chapter 2 deals with the synthesis of membranes, the properties of membranes, and the application of membranes. The beginning also identifies the three types of membranes (i.e., biological, synthetic, and theoretical) and their applications.

The details of AFM are discussed in Chap. 3. It is divided into two parts. In the introduction, a brief history of the development of AFM is given, followed by a list of manufacturers and their products. The second part contains the details of the AFM components and the experimental protocols for different AFM modes, i.e., contact, non-contact, and tapping. As the synthetic polymers are soft, generally tapping mode is preferred to study the polymeric membranes. Details of the AFM image analysis, in conjunction with synthetic polymeric membranes, are also given.

The fourth chapter examines the nodular structure of the membrane surface observed under AFM. It has been known for a long time that macromolecules form nodules at the membrane surface, and the size and the shape of the nodules strongly govern the membrane performance. In conjunction with an advanced technique such as plasma etching, AFM can reveal the nodular structure at the membrane surface more clearly than any other technique. In this chapter, the relationship between the nodular structures and the membrane preparation conditions is discussed for flat sheet membranes in the first part and hollow fibers in the second part. This chapter also deals with the roughness at the membrane surface.

Chapter 5 explores the pore structure of the membrane. Pores are clearly observed under AFM when their sizes are more than 1 nm. The method to characterize the membrane surface by the mean pore size and the pore size distribution is described, and the results are compared with those obtained from other more conventional methods.

Unlike Chaps. 4 and 5, Chap. 6 deals with the cross-sectional view of the membrane when observed under AFM. Since the technique of capturing cross-sectional views was developed only recently, relatively few images are currently available. However, this technique may have a strong influence on future research, particularly in studying cell growth under the membrane surface and fouling by blocking membrane pores.

Chapter 7 discusses the use of AFM to investigate the adhesion of particles to polymer surfaces. Adhesion of particles on membrane surfaces is the main cause of fouling. In the beginning of the chapter, a short note on DLVO (Derjaguin, Landau, Verwey, and Overbeek) theory (a theory of the stability of colloidal dispersions) has been given. However, few studies of adhesion in the membrane field by AFM have been reported. The pioneer work of Bowen's school has been described.

Finally, in Chap. 8, attempts are made to correlate the AFM parameters, such as nodule and pore sizes, to the membrane performance data. Membranes used for a variety of membrane processes, including reverse osmosis, nanofiltration, ultrafiltration, microfiltration, gas and vapor separation, pervaporation, and other membrane separation processes, are covered in this chapter. AFM parameters are also correlated to membrane biofouling. This chapter also includes applications of AFM to characterize biomedical materials, including artificial organs and drug release.

Thus, the book covers all aspects of AFM studies on the characterization of synthetic polymeric membranes. The authors believe that this book is the first attempt to find cause and effect relationships using AFM between membrane preparation, membrane characterization, and membrane performance for synthetic polymeric membranes applied in various separation processes. The authors also believe that the knowledge provided in this book will contribute to the design and preparation of improved synthetic polymeric membranes.

Importance of This Book

Although several books have already been published on AFM, they were written for different applications. The novel feature of this book is that it is focused on the study of synthetic membranes and their surfaces by AFM. For this reason, this book is monumental in the fields of both AFM and synthetic membranes. Another feature of this book is that it will provide a very useful guide to readers who wish to enter this field of study. By going through the chapters that deal with various AFM images, the reader will be exposed to the latest research results of the field. However, the strength of the book lies in its friendliness to the reader in describing details of AFM experimental methods and interpretation of experimental data, particularly when AFM is used to study membrane surfaces. Hence, the potential readers of the book are academic researchers who are investigating synthetic membranes and also R&D specialists who wish to improve and control the quality of synthetic membranes for various purposes.

This book may also attract a wider range of readers, since synthetic membranes are now considered to be one of the most important tools in the areas of seawater desalination, wastewater treatment, water production, food processing, treatment of pharmaceutical products, air and water cleaning, separation of chemical and petrochemical products, drug release, and other biomedical applications.

Ottawa, September 2007

Kailash C. Khulbe
C.Y. Feng
Takeshi Matsuura

Abbreviations and Symbols

Δt	Temperature difference
λ	Heat of vaporization of the solvent (kJ kg – mol^{-1})
λ'	Ratio of solute radius and pore radius
μ	Viscosity
μ_p	Mean pore size
μm	Micron, micrometer
ρ	Density
σ_p	Standard deviation
Å	Angstrom
AFM	Atomic force microscopy
AFS	Atomic force spectroscopy
BSA	Bovine serum albumin
°C	Degree centigrade
CA	Cellulose acetate
CAB	Cellulose acetate butyrate
C-AFM	Contact mode atomic force microscopy
cAMP	Adenosine 3′,5′-cyclic monophosphate
CE	Cellulose
cm	Centimeter
CTA	Cellulose triacetate
C_p	Heat capacity
Da	Dalton
DEHPA	Di-2-ethylhexylphosphoric acid
DLVO	DLVO (Derjaguin, Landau, Verwey, and Overbeek) theory
DMAc	*N*, *N*-Dimethylacetamide
DMF	Dimethylformamide
DSC	Differential scanning calorimetry
DSPM	Donnan-steric-pore model
DTPA	Di-2-ethylhexylthiophosphoric acid
ED	Electrodialysis
F	Force
f_{NaCl}	Apparent rejection of NaCl (%)
FE-SEM	Field emission scanning electron microscopy
FFT	Fast Fourier transform
G	Gravity constant

GBL	γ-Butyrolactone
gm	Gram
gm L^{-1}	Gram per liter
h	Hour
h	Thickness of the cast film
HEMA	Hydroxyethyl methacrylate
HFP	Hexafluoropropylene
HMDSO	Hexamethyldisiloxane
HPC	Hydroxypropylcellulose
J_v	Permeate flux (kg m^{-2} h^{-1})
J_w	Water permeate flux (kg m^{-2} h^{-1})
k	Spring constant
Kc	Thermal conductivity
kDa	Kilodalton
LEPw	Liquid entry pressure of water
LFM	Lateral force microscopy
Ma	Marangoni number
Mac	Critical Marangoni number
MD	Membrane distillation
MF	Microfiltration
min	Minute
mL	Milliliter
MPa	Megapascal
MPD	*m*-Phenylene diamine
M_w	Molecular weight
MWCO	Molecular weight cutoff
nN	Nanonewton
NC-AFM	Non-contact mode atomic force microscopy
NF	Nanofiltration
nm	Nanometer
NMMO	*N*-Methylmorpholine-*N*-oxide
NMP	*N*-Methyl-2-pyrrolidone
NMR	Nuclear magnetic resonance
PA	Polyamide
PAN	Polyacrylonitrile
PC	Polycarbonate
PE	Polyethylene
PEEK	Poly(ether ether ketone)
PEG	Polyethylene glycol
PEI	Poly(etherimide)
PES	Poly(ether sulfone)
PET	Poly(ethylene terephthalate)
PI	Polyimide
pMDA	Pyromellitic dianhydride

PMMA	Polymethyl methacrylate
ppm	Part per million
PP	Polypropylene
PAA	Polyacrylic acid
PPO	Poly(phenylene oxide) or poly(2,6-dimethyl-1,4-phenylene oxide)
PPO-C_6H_6	PPO membrane cast by PPO solution in benzene
PPO-C_6H_5Br	PPO membrane cast by PPO solution in bromobenzene
PPO-$C_6H_5CH_3$	PPO membrane cast by PPO solution in toluene
PPO-C_6H_5Cl	PPO membrane cast by PPO solution in chlorobenzene
PPO-CS_2	membrane cast by PPO solution in carbon disulfide
PPO-TCE	PPO membrane cast by PPO solution in trichloroethylene
PSf	Polysulfone
psi	Pound per square inch
PTFE	Polytetrafluoroethylene
PV	Pervaporation
PVA	Poly(vinyl alcohol)
PVDF	Poly(vinylidene fluoride)
PVP	Poly(vinylpyrrolidone)
PWP	Pure water permeation
r_p	Pore radii
r_s	Solute radii
R	Gas constant
R'	Radius of sphere
R_a	Mean roughness
Ra	Rayleigh number
Rac	Critical Rayleigh number
rms	Root mean square
RO	Reverse osmosis
s	Second
SDS	Sodium dodecyl sulfate
SEM	Scanning electron microscopy
SMMs	Surface modifying macromolecules
SPEEK	Sulfonated poly(ether ether ketone)
SPM	Scanning probe microscopy
SPPO	Sulfonated poly(phenylene oxide)
SPPOH	Sulfonated poly(phenylene oxide) in hydrogen form
SPS	Sulfonated polysulfone
STM	Scanning tunneling microscopy
T	Temperature
T_g	Transition temperature
TBP	Tributyl phosphate
TCE	Trichloroethylene
TEM	Transmission electron microscopy
TFC	Thin film composite

TFE	Tetrafluoroethylene
TIPS	Thermally induced phase separation
TM-AFM	Tapping mode atomic force microscopy
TMC	Trimesoyl chloride
TRIM	Trimethyl propane trimethacrylate
TTD	2,2,4-Trifluoro-5-trifluoromethoxy-1,3-dioxole
UF	Ultrafiltration
UV	Ultraviolet
V	Volt
VOCs	Volatile organic compounds
WAXS	Wide angle X-ray spectroscopy
wt.%	Weight percentage
XPS	X-ray photoelectron spectroscopy
Z	Difference between the highest and the lowest point within the given area (nm)

Table of Contents

1 Introduction

Thomas Graham was the father of membrane science, and he performed the first recorded experiments on the transport of gases and vapors in polymeric membranes. In 1829, he observed that a wet pig bladder inflated to the bursting point when placed in an atmosphere of carbon dioxide [1]. In 1861, Graham reported his first dialysis experiment using a synthetic membrane [2]. He also tested a permeability rate measuring device using flat membranes with a vacuum on one side, displacing a mercury column, and postulated a mechanism for the permeation process [3]. Mitchell [4,5] was the first who reported gas permeation through natural rubbers. Schoenbein [6] was the first to study cellulose nitrate, the first synthetic (or semisynthetic) polymer. Fick [7] used cellulose nitrate membranes in his classic study "Ueber Diffusion". Lord Rayleigh [8]] was the first to determine the relative permeabilities of oxygen, nitrogen, and argon in rubber. Later on, polymer membranes were used for the separation of gases, etc. [9,10]. Since the early 1960s, synthetic membranes have been used successfully in a wide variety of industrial applications.

The exact definition of a membrane is complicated, but according to Mulder [11], a general definition could be *a selective barrier between two phases, the term selective being inherent to a membrane or a membrane process*. However, the definition says nothing about membrane structure or membrane function. Membrane science arbitrarily can be divided into seven categories: material selection, material characterization and evaluation, membrane preparation, membrane characterization and evaluation, membrane transport phenomena, membrane module design, and process performance [12].

The membrane can be a solid, a liquid, or a gel, and the bulk phases can be liquid, gas, or vapor. Membranes can be classified according to their structures. Homogeneous or symmetric membranes each have a structure that is the same across the thickness of the membrane. These membranes can be porous or have a rather dense uniform structure. Heterogeneous or asymmetric membranes can be categorized into three basic structures: (1) integrally skinned asymmetric membrane with a porous skin layer, (2) integrally skinned asymmetric membrane with a dense skin layer, and (3) thin film composite membranes [13]. Porous asymmetric membranes are made by the phase inversion process [14,15] and are applied in dialysis, ultrafiltration, and microfiltration, whereas integrally skinned asymmetric membranes with a dense skin layer are applied in reverse osmosis and gas separation applications.

Thin film composite membranes consist of a thin, selective polymer layer atop a porous support. In this membrane type, the separation and mechanical functions are assigned to different layers in the membrane. This membrane type was originally developed for reverse osmosis applications; however, thin film composite membranes are also used in nanofiltration, gas separation, and pervaporation.

Membranes can be fabricated from a wide variety of organic (e.g., polymers, liquid) or inorganic (e.g., carbons, zeolite, etc.) materials. The majority of commercial membranes are made of polymers. The properties of the membrane are controlled by its material and structure.

Membranes can be made in the form of flat sheets, can be tubular, or can be made of hollow fibers and nanofibers. The development of efficient membranes depends on the knowledge of active skin morphology. The control of the polymer morphology in the selective layer is very crucial for the design of a synthetic polymeric membrane. Many attempts have been made during the past 35 years to establish the cause and effect relationship between membrane preparation, polymer morphology, and membrane performance. Although all these attempts were valuable in shedding some light on the mechanism of membrane formation and membrane transport, the understanding of the phenomena seems insufficient, mainly due to the complex nature of the mechanism.

The polymeric membrane has three important structural levels: (1) the molecular, which is equivalent to the chemical nature of the polymer, is characterized by polar, steric, and ionic factors, and is also responsible for the membrane's microcrystalline nature; (2) the microcrystalline, which affects both the transport and mechanical properties of the membrane; and (3) the colloidal, which is concerned with the aggregation of macromolecules and governs the statistics of pores (size, size distribution, density, and void volume). It is desirable to develop new characterization methods at each level to achieve a more rigorous understanding of the polymeric structure in the membrane.

Different approaches can be used to characterize the membranes, and there are various well-established methods for such characterization. There are also newly developed methods, especially for surface morphology. Standard methods for the investigation of membranes are scanning electron microscopy (SEM) [16], scanning force microscopy [17], and atomic force microscopy (AFM) [13]. Among these, AFM allows the surface study of non-conducting materials down to the scale of nanometers. It was invented by Binning et al. [18], and its main advantage over electron microscopy techniques is that no previous preparation of a sample is needed [19]. Its application to membranes, both biological and synthetic, is growing rapidly.

AFM offers a very wide range of applications and is used to solve processing and materials problems in a large range of technologies in the electronics, telecommunications, biomedical, chemical, automotive, aerospace, and energy industries. Materials that can be studied include thin and thick film coatings, ceramics, composites, glasses, synthetic and biological membranes, metals, polymers, and semiconductors. AFM is also used to study phenomena such as abrasion, adhesion, cleaning, corrosion, etching, friction, lubrication, plating, and polishing. AFM images show critical information about surface features with unprecedented clarity and can examine any rigid surface. Minor (and major) differences between "smooth" surfaces are shown dramatically. AFM can resolve very tiny features, even single atoms, that were previously unseen. It can examine a field of view larger than 125 μm (0.005 in), so as to make comparisons with other information, e.g., features seen in the light microscope or as seen by eye. AFM can also examine rough surfaces, since its vertical range is more than 5 μm. It can achieve a resolution of 100 pm, and unlike electron microscopes, can image samples in air and under liquids. AFM was first applied to polymer surfaces in 1988 [20], shortly after its invention [18]. It is frequently applied to polymer surfaces, principally to reveal surface morphology, nanostructure, chain packing, and conformation. Hansma et al. [21] studied molecular resolution images of a nonconductive organic monolayer and an amino acid crystal that revealed individual methyl groups on the ends of the amino acids. AFM may be used to quantify the three parameters that most influence membrane separation performance: pore size distribution, membrane surface electrical properties, and membrane adhesion (fouling). Currently, AFM is becoming a very important tool for the characterization of synthetic membranes. Adhesion, attraction, and repulsion between surfaces in liquids can be studied [22].

References

1. Graham T (1829) Roy Inst J
2. Graham T (1861) Phil Trans R Soc 151:183
3. Graham T (1866) Philos Mag 32:401
4. Mitchell JK (1831) Roy Inst J 2:101
5. Mitchell JK (1831) Roy Inst J 2:307
6. Schoenbein C (1846) British Patent 11 402
7. Fick A (1855) Ann Phys Chem 94:59
8. Lord Rayleigh JW (1900) Philos Mag 49:220
9. Matthes A (1944) Kolloid Z 108:79
10. Barrer RM, Barrie JA, Slater J (1958) J Polym Sci 27:177
11. Mulder M (1996) Basic principles of membrane technology. Kluwer, Dordrecht
12. Lloyd DR (1985) Membrane materials science: an overview. In: Lloyd DR (ed) Materials science of synthetic membranes. ACS Symposium series 269. American Chemical Society, Washington, DC, p 1
13. Pinnau I, Freeman BD (1999) Formation and modification of polymeric membranes: overview. In: Pinnau I, Freeman BD (eds) Membrane formation and modification, ACS symposium 744. American Chemical Society, Washington, DC, p 1
14. Kesting RE (1971) Synthetic polymeric membranes. McGraw-Hill, New York
15. Strathmann H (1985) In: Porter MC (ed) Handbook of industrial membrane technology. Noyes, Park Ridge, p 1
16. Kim KJ, Fane AG (1994) J Membr Sci 88:103
17. Magonov SN, Wangbo MH (1996) Surface analysis with STM and AFM: experimental and theoretical aspects of image analysis. Wiley-VCH, Weinheim
18. Binning G, Quate CF, Gerber CH (1986) Phys Rev Lett 56:930
19. Nakao S (1994) J Membr Sci 96:131
20. Albrecht TR, Dovek MM, Lang CA, Grutter P, Quate CF, Kuan SNJ, Frank CW, Pease RFW (1988) J Appl Phys 64:1178
21. Hansma PK, Elings VB, Marti O, Bracker CE (1988) Science 242:209
22. Weisenhorn AL, Maivald P, Butt HJ, Hansma PK (1992) Phys Rev B Condens Matter Mater Phys 45:11226

2 Synthetic Membranes for Membrane Processes

2.1 Introduction

According to Wikipedia [1], a membrane is a thin, typically planar structure or material that separates two environments or phases and has a finite volume. It can be referred to as an interphase rather than an interface. Membranes selectively control mass transport between phases or environments. Again, according to Wikipedia, membranes can be divided into three groups: (1) biological membranes, (2) artificial membranes, and (3) theoretical membranes.

Biological membranes include:

1. Cell membranes and intracellular membranes
2. Mucous membranes
3. S-layer
4. Serous membranes and mesothelia that surround organs, including:
   a) The peritoneum that lines the abdominal cavity
   b) The pericardium that surrounds the heart
   c) The pleura that surrounds the lungs
   d) The periosteum that surrounds bone
   e) The meninges that surround the brain (the dura mater, the arachnoid, and the pia mater)

Artificial membranes are used in:

1. Reverse osmosis
2. Filtration (microfiltration, ultrafiltration)
3. Pervaporation
4. Dialysis
5. Emulsion liquid membranes
6. Membrane-based solvent extraction
7. Membrane reactors
8. Gas permeation
9. Supported liquid membranes

This book is devoted to synthetic, or artificial, membranes. In particular, our focus will be on polymeric synthetic membranes, since most industrial membranes belong to this category. Before entering the main subject of this book, i.e., atomic force microscopy, the current status of synthetic polymeric membranes is outlined. Thus, in the following pages, we will provide some information about the preparation of membranes, the properties of membranes, and their applications.

2.2 Membrane Preparation

Synthetic membranes are fabricated in two main geometries:

1. Flat sheet—utilized in the construction of flat sheet, disc, spirally wound, plate, and frame modules
2. Cylindrical—utilized in tubular and capillary, or hollow fiber, modules

Membranes can be prepared from both ceramic and polymeric materials. Ceramic materials have several advantages over polymeric materials, such as higher chemical and thermal stability. However, the market share of polymeric membranes is far greater than ceramic membranes as the polymeric materials are easier to process and less expensive. A handful of technical polymers are currently used as membrane materials for 95% of all practical applications [2]. Polymeric materials that are used to prepare separation membranes are mostly organic compounds. A number of different techniques are available to prepare synthetic membranes.

2.2.1 Membranes with Symmetric Structure

Although most of the practically useful membranes are asymmetric, as explained later, some of the membranes have symmetric structures. They are prepared in the following ways:

Track etching A sheet of polymeric film moves underneath a radiation source and is irradiated by high-energy particles. The spots that are subjected to bombardment of the particles are degraded or chemically altered during this process. Then, the film undergoes an etching process in an alkaline or hydrogen peroxide bath (depending on the material), where the polymer is etched along the path of high-energy particles.

Precipitation from the vapor phase A cast polymer solution that consists of polymer and solvent is brought into a nonsolvent vapor environment saturated with solvent vapor. The saturated solvent vapor suppresses the evaporation of solvent from the film; the nonsolvent molecules diffuse into the film causing polymer coagulation.

2.2.2 Membranes with Asymmetric Structure

Most membranes used in industries have an asymmetric structure. Figure 2.1 shows schematically a typical cross-sectional view of an asymmetric membrane [3]. It consists of two layers: the top one is a very thin dense layer (also called the top skin layer), and the bottom one is a porous sublayer. The top dense layer governs the performance (permeation properties) of the membrane; the porous sublayer only provides mechanical strength to the membrane. The membranes of symmetric structures do not possess a top dense layer. In the asymmetric membrane, when the material of the top

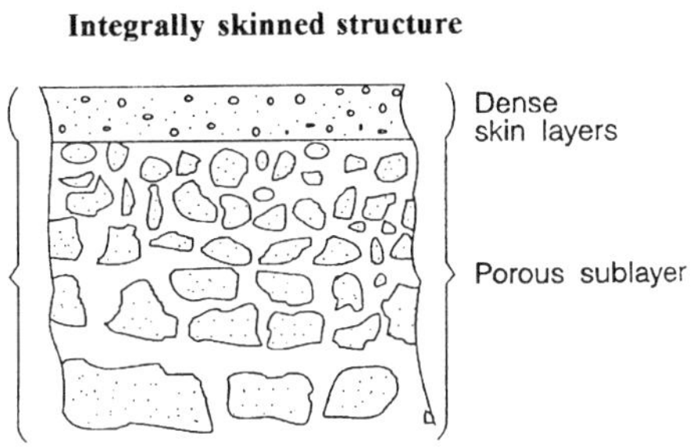

Fig. 2.1. Cross-sectional view of an asymmetric membrane. Reprinted from [3], with kind permission from the author

layer and porous sublayer are the same, the membrane is called an *integrally skinned* asymmetric membrane. On the other hand, if the polymer of the top skin layer is different from the polymer of the porous sublayer, the membrane is called a *composite* membrane. The advantage of the composite membrane over the integrally skinned asymmetric membrane is that the material for the top skin layer and the porous sublayer can be chosen separately to optimize the overall performance. There are various methods for the preparation of asymmetric membranes, which are described in the sections that follow.

2.2.2.1 Phase Inversion Technique for Preparation of Integrally Skinned Asymmetric Membranes

Dry–wet phase inversion technique (Loeb-Sourirajan method) A number of methods can be used to achieve phase inversion. Among these, the dry–wet phase inversion technique and thermally induced phase separation (TIPS) are the most commonly used in membrane manufacturing. The dry–wet phase inversion technique, also called the *Loeb-Sourirajan* technique, was used by Loeb and Sourirajan in their development of the first cellulose acetate membrane for seawater desalination [4]. In this method, a polymer solution is prepared by mixing polymer and solvent (sometimes even nonsolvent). The solution is then cast on a suitable surface by a *doctor blade* to a precalculated thickness. After a partial evaporation of the solvent, the cast film is immersed in a nonsolvent medium called a *gelation bath*. Due to a sequence of two desolvation steps, i.e., evaporation of the solvent and solvent–nonsolvent exchange in the gelation bath, solidification of the polymer film takes place. It is desirable to choose a solvent of strong dissolving power with a high volatility. During the first step of desolvation by solvent evaporation, a thin skin layer of solid polymer is formed instantly at the top of the cast film due to the loss of solvent. In the solvent–nonsolvent exchange process that follows, the nonsolvent diffuses into the polymer solution film through the thin solid layer while the solvent diffuses out. The change in the composition of the polymer solution film during the solvent–nonsolvent exchange process, often called a composition path, is illustrated schematically in Fig. 2.2 (lines A, B, and C each represent a composition path).

The top skin layer can also be made porous by lowering the polymer concentration in the casting solution and the solvent evaporation period. This is called, hereafter, the porous skin layer. Asymmetric membranes can also be made in a tubular form using a casting bob assembly and a hollow fiber spinneret [5].

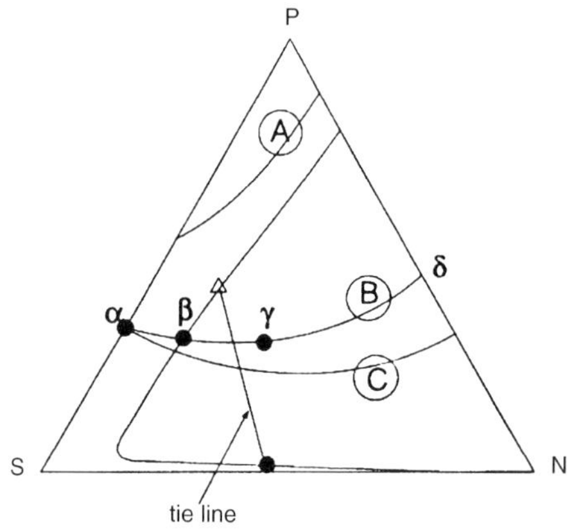

Fig. 2.2. Triangular diagram of polymer (*P*), solvent (*S*), and nonsolvent (*N*). Reprinted from [3], with kind permission from the author

Thermally induced phase separation method In this method, phase inversion is introduced by lowering the temperature of the polymer solution. A polymer is mixed with a substance that acts as a solvent at a high temperature and the polymer solution is cast into a film. When the solution is cooled, it enters into an immiscible region due to the loss of solvent power. Because the solvent is usually nonvolatile, it must be removed with a liquid that is miscible with the solvent but not miscible with the polymer.

2.2.2.2 Preparation of Composite Membranes

Dip coating An integrally skinned asymmetric membrane with a porous skin layer (called hereafter a substrate membrane) is prepared from a polymer solution by applying the dry–wet phase inversion method. The membrane is then dried according to the method described later, before it is dipped into a bath containing a dilute solution of another polymer. When the membrane is taken out of the bath, a thin layer of coating solution is deposited on the top of the substrate membrane. The solvent is then removed by evaporation, leaving a thin layer of the latter polymer on top of the substrate membrane.

Interfacial polymerization This method, developed by Cadotte and the coworkers of Film Tech in the 1970s, is currently most widely used to prepare high performance reverse osmosis and nanofiltration membranes [6]. A thin selective layer is deposited on top of a porous substrate membrane by interfacial in situ polycondensation. There are a number of modifications of this method primarily based on the choice of the monomers [7]. However, for simplicity, the polycondensation procedure is described by a pair of diamine and diacid chloride monomers.

A diamine solution in water and a diacid chloride solution in hexane are prepared. A porous substrate membrane is then dipped into the aqueous solution of diamine. The pores at the top of the porous substrate membrane are filled with the aqueous solution in this process. The membrane is then immersed in the diacid chloride solution in hexane. Since water and hexane are not miscible, an interface is formed at the boundary of the two phases. Polycondensation of diamine and diacid chloride

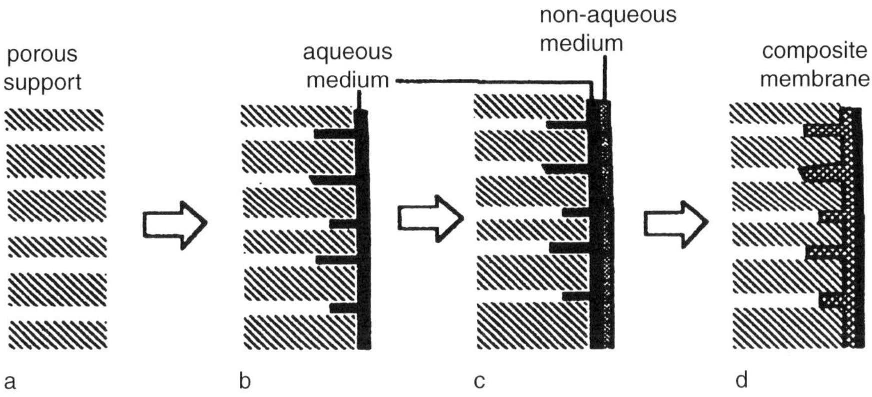

Fig. 2.3. Steps in the formation of a composite membrane via interfacial polymerization. Reprinted from [3], with kind permission from the author

takes place at the interface, resulting in a very thin layer of polyamide. The preparation of composite membranes by interfacial in situ polycondensation is schematically presented in Fig. 2.3.

2.2.2.3 Membrane Surface Modification

As mentioned above, the top skin layer governs the performance of a separation membrane. The surface deposition of contaminants from solutions or from gas mixtures is also affected by the surface properties of the membrane. This is particularly important when decline in the membrane flux with a prolonged operating period is observed, since it is often caused by the contaminant deposition. Hence, many attempts have been made to modify the membrane surface, aimed at prevention of contaminant deposition and maintenance of high flux. Several methods of surface modification are described below.

Chemical modification The surface of a membrane can be modified by chemical reactions. For example, when the surface of a polyamide composite membrane is brought into contact with a strong hydrofluoric acid solution, the top polyamide layer becomes slightly thinner by a chemical reaction with hydrofluoric acid. As a result, the flux increases considerably while the rejection of sodium chloride is unchanged or slightly increased [8].

Plasma polymerization When a vacuum is maintained inside a tubular reactor and a high frequency electric field is applied outside, a glow discharge is generated inside the reactor (Fig. 2.4). Plasma that consists of various ions, radicals, electrons, and molecules is formed in the glow discharge. When a porous substrate membrane is placed in the plasma, the surface of the membrane is subjected to various changes corresponding to the property of plasma. The substrate surface can be etched and/or chemically active sites can be introduced to the surface, and, upon contact with organic compounds, an irregular polymerization can occur at the substrate surface. This is called plasma polymerization [9].

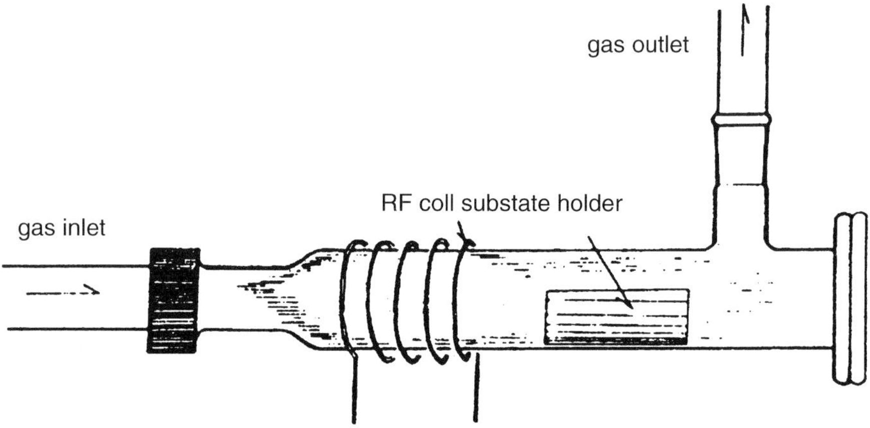

Fig. 2.4. Reactor for plasma treatment. Reprinted from [3], with kind permission from the author

Graft polymerization The surface of a porous substrate membrane is irradiated with γ-rays, which causes the generation of radicals on the membrane surface. Then, the membrane is immersed in a monomer solution. The graft polymerization of the monomers is initiated at the membrane surface. By choosing a very hydrophilic monomer, the hydrophilicity of the surface is increased considerably.

Surface modification by surface modifying macromolecules (SMMs) In a polymer blend, thermodynamic incompatibility between polymers usually causes demixing of polymers. If the polymer is equilibrated in air, the polymer with the lowest surface energy (the hydrophobic polymer) will concentrate at the air interface and reduce the system's interfacial tension as a consequence. The preferential adsorption of a polymer of lower surface tension at the surface was confirmed by a number of researchers for the miscible blend of two different polymers. Based on this concept, surface modifying macromolecules as surface-active additives were synthesized and blended into polymer solutions of poly(ether sulfone) (PES). Depending on the hydrophobic [10, 11] or hydrophilic [12] nature of the SMMs, the membrane surface becomes either more hydrophobic or more hydrophilic than the base polymeric material.

2.2.3 Membrane Drying

The wet cellulose acetate membranes prepared for reverse osmosis purposes can be used for gas separation when they are dried. The water in the cellulose acetate membrane cannot be evaporated in air, however, since the asymmetric structure of the membrane will collapse. Instead, the multi-stage solvent exchange and evaporation method is applied. In this method, a water-miscible solvent such as ethanol first replaces the water in the membrane. Then, a second volatile solvent such as hexane replaces the first solvent. The second solvent is subsequently air-evaporated to obtain a dry membrane [13, 14]. The reason for replacing water with hexane is to reduce the capillary force inside the pore so that it will not collapse during the drying process.

2.3 Membranes for Separation Processes

2.3.1 Membranes for the Separation of Solutions and Solvent Mixtures

Membranes for the separation of solutions and liquid mixtures may be distinguished on the basis of pore sizes as reverse osmosis (RO, below 1 nm), ultrafiltration (UF, 2–100 nm), and microfiltration (MF, 100 nm to 2 μm), although this classification is very arbitrary. Pore sizes of nanofiltration (NF) membranes are between RO and UF membranes.

2.3.1.1 Reverse Osmosis Membranes

An RO membrane acts as a barrier to flow, allowing selective passage of a particular species (solvent) while other species (solutes) are retained partially or completely. Solute separation and permeate solvent (water in most cases) flux depend on the material selection, the preparation procedures, and the structure of the membrane barrier layer [5, 15]. Cellulose acetate (CA) is the material for the first generation reverse osmosis membrane. The announcement of CA membranes for sea water desalination by Loeb and Sourirajan in 1960 triggered the applications of membrane separation processes in many industrial sectors. CA membranes are prepared by the dry–wet phase inversion technique. Another polymeric material for RO is aromatic polyamide [16].

In aromatic polyamide polymers, aromatic rings are connected by an amide linkage, -CONH–. While the aromatic ring attached to –NH– is metasubstituted, the ring attached to -CO– is the mixture of meta- and parasubstitutions, which gives more flexibility to the polymeric material. Aromatic polyamide remains one of the most important materials for reverse osmosis membranes since the thin selective layer of composite membranes is aromatic polyamide synthesized by interfacial in situ polymerization.

2.3.1.2 Nanofiltration Membranes

Most NF membranes are negatively charged. In interfacial polycondensation, trimesoyl (triacid) chloride is often mixed with phthaloyl (diacid) chloride in the acidic component of the polycondensation reaction. Although most carboxylic groups are consumed to form amide linkage, a small portion of the carboxylic groups do not participate in the reaction, becoming the source of the electric charge. Since -COOH becomes $-COO^-$ upon dissociation, the membranes are negatively charged. Because of the negative charge, anions are preferentially rejected by nanofiltration membranes.

Another method of preparing nanofiltration membranes is to dip-coat a thin layer of sulfonated poly(phenylene oxide) (SPPO) [17], sulfonated polysulfone (SPS) [18], or carboxylated polysulfone [19] on a porous substrate membrane. The sulfonic acid groups in SPPO and SPS also become negatively charged with $-SO_3^-$ groups upon dissociation. Sulfonic acid is a stronger acid than carboxylic acid.

2.3.1.3 Ultrafiltration Membranes

Ultrafiltration is primarily a size-exclusion-based, pressure-driven membrane separation process. UF membranes typically have pore sizes in the range of 2–500 nm and retain species in the molecular range from 3000 to 500 000 Da [20], while solvent (water) passes through the membrane. UF membranes have a porous skin layer. The most important UF membrane properties are the membrane productivity (flux) and extent of separation (rejection of various feed components).

In contrast to the polymeric materials for reverse osmosis and nanofiltration membranes, for which the macromolecular structures have much to do with permeation properties such as salt rejection characteristics, the choice of membrane material for ultrafiltration does not depend on the material's influence on the permeation properties.

Membrane permeation properties are largely governed by the pore sizes and the pore size distributions of UF membranes. Rather, thermal, chemical, mechanical, and biological stability are considered of greater importance. Typical UF membrane materials are polysulfone (PS), poly(ether sulfone), poly(ether ether ketone) (PEEK), cellulose acetate and other cellulose esters, polyacrylonitrile (PAN), poly(vinylidene fluoride) (PVDF), polyimide (PI), poly(etherimide) (PEI), and aliphatic polyamide (PA). All these polymers have a T_g higher than 145 °C except for cellulose esters. They are also stable chemically and mechanically, and their biodegradability is low. The membranes are made by the dry–wet phase inversion technique.

2.3.1.4 Microfiltration Membranes

Polymeric materials for MF membranes cover a very wide range, from relatively hydrophilic to very hydrophobic materials. Typical hydrophilic materials are polysulfone, poly(ether sulfone), cellulose (CE) and cellulose acetate, polyamide, polyimide, poly(etherimide) and polycarbonate (PC). Typical hydrophobic materials are polyethylene (PE), polypropylene (PP), polytetrafluoroethylene (PTFE, Teflon) and poly(vinylidene fluoride).

Hydrophilic MF membranes can be made by the dry–wet phase inversion technique, which can also be used to make PVDF membranes. On the other hand, other hydrophobic microfiltration membranes are made by the thermally induced phase separation technique. In particular, semicrystalline PE, PP, and PTFE are stretched parallel to the direction of film extrusion so that the crystalline regions are aligned in the direction of stretch, while the noncrystalline region is ruptured, forming long and narrow pores. Hydrophobic membranes do not allow penetration of water into the pore until the transmembrane pressure drop reaches a threshold called the liquid entry pressure of water (LEPw). These membranes can therefore be used for membrane distillation. The track-etching method is applied to make microfiltration membranes from PC.

An especially important characteristic of a microfiltration membrane is uniform pores with as many of them per unit area as possible, and with the thinnest possible layer where these pores are at their smallest. The use of MF membranes is the quantitative separation of suspended matter in the 0.1–10 μm size range from liquids and gases.

2.3.2 Membranes for Gas and Vapor Separation

The concept of separating gases with polymeric membranes is more than 100 years old, but the widespread use of gas separation membranes has occurred only within the last 20–30 years. Separation is achieved because of differences in the relative transport rates of feed components. Components that diffuse more rapidly become enriched in the low pressure *permeate* stream, while the slower components are concentrated in the *retentate*, or residue, stream. The membrane process that separates components based on their relative rates of permeation distinguishes it from equilibrium processes such as distillation or extraction.

Gas and vapor separation membranes are classified into two categories. In the first, rubbery polymers such as silicone rubber, natural rubber, and poly(4-methyl-1-pentene) are used to take advantage of their high permeabilities, even though selectivities are rather moderate. Production of enriched oxygen for medical purposes is performed by this type of membrane with an oxygen/nitrogen selectivity of about two. Asymmetric membranes made from glassy polymers such as cellulose acetate and other cellulose derivatives, polycarbonate, aromatic polyamide, aromatic polyimides, and poly(phenylene oxide) (PPO) and its derivatives belong to the second category.

These asymmetric membranes are made by the dry–wet phase inversion technique. Membranes must be dried before being used. Solvent exchange is necessary to dry cellulose acetate membranes. These membranes take advantage of the high selectivity of glassy polymers. The selective dense layer at the top of the membrane must be very thin so that a high flux can be achieved. They are used in a wide range of industrial gas separation processes such as hydrogen recovery from various chemical syntheses, sour gas removal from natural gas and production of nitrogen-enriched air. For the asymmetric membranes to be effective in gas separation, the thin selective layer at the top of the membrane should be perfect. This requirement is more stringent in gas separation membranes than liquid separation membranes since defective pores cannot be automatically closed when the surface is in contact with dry gas. In contrast, defective pores of RO and pervaporation (PV) membranes can be closed by the swelling of the top skin layer when it is brought into contact with feed liquid.

Since it is difficult to make a selective skin layer defect-free, a method was proposed by Henis and Tripodi to seal defective pores. Their method was applied to asymmetric polysulfone membranes, which led to the production of the commercial Prism membrane [21].

According to the method, a relatively thick silicone rubber layer is coated on a thin selective layer of an asymmetric polysulfone membrane. The thickness of silicone rubber is about 1 μm while the effective thickness of the selective polysulfone layer is one tenth of 1 μm. While being coated, silicone rubber penetrates into the pores to plug them. Thus, feed gas is not allowed to leak through the defective pores. The selectivity of the membrane approaches that of the defect-free polysulfone layer. Moreover, since the permeabilities of silicone rubber for gases are orders of magnitudes higher than those of polysulfone, the permeation rate is not affected very much even when a relatively thick silicone rubber layer is coated.

Membranes for vapor removal from air have a structure similar to the Prism membrane, but they are prepared on a different principle [22]. Aromatic poly(etherimide) is used to produce a porous substrate membrane by the dry–wet phase inversion method. This polymer was chosen over polysulfone/poly(ether sulfone) due to the higher durability of poly(etherimide) to organic vapors. Unlike an asymmetric polysulfone substrate for the Prism membrane, the top layer of the asymmetric poly(etherimide) membrane has a large number of pores, the size of which is equivalent to those of ultrafiltration membranes. When a layer of silicone rubber is coated on the top layer of the porous substrate membrane, the silicone rubber layer will govern the selectivity, and the porous support will provide only mechanical strength to the composite membrane. Since the permeabilities of water and organic vapors through the silicone rubber layer are much greater than those of oxygen and nitrogen, these membranes are effective in dehumidification of air and removal of organic vapors from air.

2.3.3 Membranes for Pervaporation and Membrane Distillation

Pervaporation and membrane distillation (MD) are distinguished from the above membrane separation processes since phase change, from liquid to vapor, takes place in the process.

2.3.3.1 Pervaporation

Pervaporation is characterized by the imposition of a barrier (membrane) layer between a liquid and a vaporous phase, with a mass transfer occurring selectively across the barrier to the vapor side. Separation occurs with the efficacy of the separation effect being determined by the physiochemical structure of the membrane.

Pervaporation membranes were developed for the dehydration of ethanol and other organic solvents. Therefore, the dense selective layer is made of polyvinyl alcohol that is one of the most hydrophilic materials. Water is preferentially sorbed to polyvinyl alcohol and also preferentially transported. To suppress the excessive swelling of polymer in water, polyvinyl alcohol is partially cross-linked by dialdehydes such as glutaraldehyde [23].

The dense polyvinyl alcohol layer is supported by a porous PAN substrate membrane. Polyelectrolyte material [24] and chitosan [25], a natural product, are also potentially useful for dehydration by pervaporation. Silicone rubber membranes developed for the removal of organic vapors from air can also be used for the removal of volatile organic compounds (VOCs) from water by pervaporation [23]. Because of the high hydrophobic nature of silicone rubber, VOCs are preferentially sorbed and transported through the membrane.

2.3.3.2 Membrane Distillation

Membrane distillation is similar to pervaporation since phase change is involved in the process. When feed liquid (usually water) is in contact with a nonwetted porous hydrophobic membrane, water does not enter into the pores because the feed liquid is maintained below a threshold pressure, the liquid penetration pressure of water. Only water vapor permeates through the pores from the feed to the permeate side. The driving force is the vapor pressure drop from the feed to the permeate side, since the permeate temperature is maintained below the feed temperature. Commercial hydrophobic membranes made of polypropylene, poly(vinylidene fluoride) and polytetrafluoroethylene, either in capillary or flat-sheet form, are used for MD, although these membranes were primarily prepared for microfiltration purposes. With a salt solution, for example NaCl in water, only water has a vapor pressure, i.e., the vapor pressure of NaCl can be neglected, which means that only water will permeate through the membrane, and consequently very high selectivities are obtained.

2.3.4 Membranes for Other Separation Processes

While all the above mentioned membrane separation processes utilize the transmembrane pressure drop as the driving force, there are other membrane separation processes based on different driving forces.

2.3.4.1 Electrodialysis

Membranes for electrodialysis (ED) are either positively or negatively charged. Ionic species in the solution are transported through the membrane by the electrical potential difference between the two sides of the membrane. When a membrane is positively charged, it is called an anion exchange membrane since only anions are allowed to permeate through the membrane. A negatively charged membrane is called cationic since only cations are allowed to permeate through the membrane. The base polymeric material is polystyrene cross-linked by divinylbenzene. Quaternary ammonium cations are attached to some aromatic rings of anionic membranes, while sulfonic groups or carboxylic groups are attached to some aromatic rings of cationic membranes [26].

2.3.4.2 Dialysis

Dialysis is the separation of smaller molecules from larger molecules, or dissolved substances from colloidal particles, in a solution by selective diffusion through a semipermeable membrane. Dialysis is a rate-governed membrane process in which a microsolute is driven across a semipermeable membrane by means of a concentration gradient. The microsolute diffuses through the membrane at a greater rate than macrosolutes also present in the feed solution. Ordinary dialysis is referred to as diffusion of neutral molecules. If electrolytes are separated with neutral membranes, or with charged membranes, then the *Donnan effects* arising from the unequal distribution of ions interfere with the normal dialysis process. This type of dialysis is called Donnan dialysis. In the medical field, it is the process used for cleaning blood, artificially, with special equipment. Hemodialysis membranes have ultrafiltration capacities ranging from 5 to 70 $\mathrm{mL\,h^{-1}\,m^{-2}\,mmH_g^{-1}}$. Donnan dialysis makes use of ion selective membranes to provide improved selectivity.

2.4 Membrane Applications

The major applications of membranes for membrane separation processes are summarized in Table 2.1.

Table 2.1. Applications of synthetic membranes

Membranes	Applications
Reverse osmosis	1. Sea water and brackish water desalination 2. Waste water treatment (industrial and municipal, pulp and paper, textile waste water) 3. Production of boiler quality water for steam generation 4. Petroleum industry 5. Recovery of plating chemicals from wastewaters and process waters in the electroplating and metal-finishing industry
Nanofiltration	1. Water treatment 2. Product and chemical recovery 3. Concentration/dewatering 4. Fractionation of monovalent and divalent cations 5. Water softening
Ultrafiltration	1. Electrodialysis pretreatment 2. Electrophoretic paint 3. Cheese whey treatment 4. Juice clarification 5. Recovery of textile sizing agents 6. Wine clarification 7. Separation of oil/water emulsion 8. Concentration of latex emulsion from wastewater 9. Dewaxing 10. Deasphalting 11. Egg-white preconcentration 12. Kaolin concentration 13. Water treatment 14. Affinity membranes 15. Reverse osmosis pretreatment
Microfiltration	1. Purification of fluids in semiconductor manufacturing industry 2. Clarification and biological stabilization in the beverage industry 3. Sterilization (in the food and pharmaceutical industries) 4. Analysis
Gas separation	1. Hydrogen recovery a) Synthesis gas ratio adjustment (H_2/CO) b) H_2 recovery from hydroprocessing purge streams c) H_2 recovery from ammonia plant purge streams and other petrochemical plant streams 2. Oxygen/nitrogen separation 3. Helium recovery 4. Removal of acid gases from light hydrocarbons 5. Biogas processing 6. Separation of organic vapors from air
Pervaporation	1. Removal of organics from water 2. Water removal from liquid organics 3. Organic/organic separation

Table 2.1. continued

Membranes	Applications
Vapor permeation	Removal of organics from air
Electrodialysis	1. Desalination of brackish water 2. Production of table salt 3. Waste water treatment 4. Concentration of RO brines 5. Applications in the chemical, food, and drug industries
Dialysis	1. Hemofiltration and hemodiafiltration 2. Donnan dialysis 3. Alcohol reduction of beverages

2.5 Membrane Characterization

The performance of membranes depends on their properties, which may be quantified by membrane characterization. The methods for membrane characterization are listed below.

Characterization of the bulk membrane polymer Durability of the membrane in the operational environment depends on the thermal, mechanical, and chemical properties of the membrane polymer. They are characterized by differential scanning calorimetry (DSC), tensile strength measurement by contacting the membrane with solutions, and the gases to be treated. Wide angle X-ray spectroscopy (WAXS) is also used to measure the crystallinity of the polymer, on which many other polymeric properties depend.

Characterization of the membrane surface It should be emphasized that the properties of the membrane surface strongly affect membrane performance. Contact angle is often used as a measure of surface hydrophilicity or hydrophobicity. X-ray photoelectron spectroscopy (XPS) provides the data on atomic compositions at the membrane surface. Recently, attentions have been focused on the nodular structure as well as the roughness at the membrane surface that can be measured by atomic force microscopy (AFM).

Pore size and pore size distribution It is obvious that the pore size and the pore size distribution of the membrane affect membrane performance. A number of methods can be used to determine the pore size and the pore size distribution. Conventional methods include bubble point method, mercury porometry, thermporometry, permporometry, and gas adsorption. Transport data of gases and solutions with solute probes can also be used to determine the pore size and the pore size distribution. Pores can also be observed by scanning electron microscope (SEM) and transmission electron microscope (TEM). Atomic force microscope can observe the pores only on the membrane surface.

References

1. Membrane. (2007) Wikimedia Foundation Wikipedia. http://en.wikipedia.org/wiki/Membrane
2. Peinemann K (2004) Next generation membrane materials. In: Abstracts of the 15th annual meeting of the NAMS, Honolulu, 26–30 June 2004
3. Matsuura T (1994) Synthetic membranes and membrane separation processes. CRC, Boca Raton, p 15
4. Loeb S, Sourirajan S (1963) Adv Chem Ser 38:117
5. Sourirajan S, Matsuura T (1985) Reverse osmosis/ultrafiltration process principles. National Research Council of Canada, Ottawa, p 802
6. Rozelle LT, Cadotte JE, Cobian KE, Kopp CVJr (1977) Nonpolysaccharide membrane for reverse osmosis: NS-100 membranes. In: Sourirajan S (ed) Reverse osmosis and synthetic membranes: theory, technology, engineering. National Research Council of Canada, Ottawa, p 249
7. Peterson RJ (1993) J Membr Sci 83:81
8. Kulkarni A, Mukherjee D, Gill WN (1994) Chem Eng Commun 129:53
9. Hirotsu T (1987) Ind Eng Chem Res 26:1287
10. Suk DE, Chowdhury G, Narbaitz RM, Santerre JP, Matsuura T, Glazier G, Deslandes Y (2002) Macromolecules 35:3017
11. Khayet M, Suk DE, Narbaitz RM, Santerre JP, Matsuura T (2003) J Appl Polym Sci 89:2902
12. Hester JF, Banerjee P, Won YY, Akthakul A, Acar MH, Mayes AM (2002) Macromolecules 35:7652
13. Lui A, Talbot FDF, Sourirajan S, Fouda AE, Matsuura T (1988) Sep Sci Technol 23:1839
14. Gantzel PK, Merten U (1970) Ind Eng Chem Process Des Dev 9:331
15. Lloyd D (1985) Membrane materials science: an overview. In: Lloyd DR (ed) Materials science of synthetic membranes. ACS Symposium Series 269. American Chemical Society, Washington, DC, p 1
16. Hoehn HH (1985) Aromatic polyamide membranes. In: Lloyd DR (ed) Materials science of synthetic membranes. ACS Symposium Series 269. American Chemical Society, Washington, DC, p 81
17. Matsuura T (2001) Reverse osmosis and nanofiltration by composite polyphenylene oxide membranes. In: Chowdhury G, Kruczek B, Matsuura T (eds) Polyphenylene oxide and modified polyphenylene oxide membranes. Kluwer, Dordrecht, p 181
18. Allegrezza AEJr, Parekh BS, Parise PL, Swiniarski EJ, White JL (1987) Desalination 64:285
19. Guiver MD, Tremblay AY, Tam CM (1989) Reverse osmosis membrane from novel hydrophilic polysulfone. In: Sourirajan S, Matsuura T (eds) Advances in reverse osmosis and ultrafiltration. National Research Council of Canada, Ottawa, p 53
20. Kulkarni SS, Funk EW, Li N (1992) Ultrafiltration. In: Ho WSW, Sirkar KK (eds) Membrane handbook. Van Nostrand, New York, p 393
21. Henis JMS, Tripodi MK (1981) J Membr Sci 7:233
22. Behling RD, Ohlrogge K, Peinemann KV (1989) The separation of hydrocarbons from waste vapor streams. In: Fouda AE, Hazlett JD, Matsuura T, Johnson J (eds) Membrane separations in chemical engineering. AIChE Symposium Series 272, New York, p 68
23. Koops GH, Smolders CA (1991) Estimation and evaluation of polymeric materials for pervaporation membranes. In: Huang RYM (ed) Pervaporation membrane separation processes. Elsevier, New York, p 253
24. Tsuyumoto M, Karakane H, Maeda Y, Tsugaya H (1991) Desalination 80:139
25. Feng XS, Huang RYM (1996) J Membr Sci 116:67
26. Strathmann H (1992) Electrodialysis. In: Ho WSW, Sirkar KK (eds) Membrane handbook. Van Nostrand, New York, p 219

3 Atomic Force Microscopy

3.1 Introduction

When we think of microscopes, we think of optical or electron. The former uses a series of glass lenses for magnification of up to ×1000. The latter creates a magnified image by focusing electrons, using magnetic fields of special coils. In the evolution of microscopy, the electron microscope improved the magnified image up to ×100 000. But both methods generate only two-dimensional images.

With the continuing evolution of the microscope comes atomic force microscopy. It can magnify up to ×1 000 000 in all three dimensions of a horizontal x, y-plane and a vertical z-plane. AFM uses a combination of the principles of the scanning tunneling microscope and the stylus profile meter. It incorporates a probe that does not damage the surface [1]. This relatively new technology is being used in the electronics, telecommunications, biomedical, chemical and membrane industries. Material currently under investigation using AFM includes thick and thin film coating, ceramics, composites, glasses, synthetic and biological membranes, metals, polymers, and semiconductors. AFM is also being used to study phenomena such as abrasion, adhesion, corrosion, cleaning, etching, plating, friction, and lubrication to name a few.

AFM can demonstrate detailed information about rigid surface features in air or immersed in liquid. Even minor differences previously unable to be seen can be distinguished via AFM. It can differentiate even single atoms in a field of view larger than 125 µm (0.005 in). This combination of exquisite detail in a three-dimensional view establishes important quantitative data analysis (such as feature sizes, surface roughness and area, and cross-section plots). Bowen et al. [2] wrote an excellent article on the atomic force microscopic studies on membranes.

In 1981, G. Binnig and H. Rohrer from the IBM research laboratory in Ruschlikon invented a new type of imaging instrument called a scanning tunneling microscope (STM) and received the Nobel Prize in 1986. Its most striking feature is the extremely high spatial resolution of the order of 0.01 nm that can be achieved, allowing one to image and even to manipulate individual atoms. The main difference between this technique and the ones mentioned earlier is that there is no need for any lenses, light or electron sources. It is the tunneling effect, a quantum mechanical property, that provides the physical foundation for this technique: simply apply a voltage between a sharp metallic tip and the investigated surface, both separated by a vacuum barrier. If this vacuum barrier is about a few atomic diameters thick, electrons are able to *tunnel* through it, and a current will flow. The current depends exponentially on a barrier distance. Hence, by scanning the tip over the surface at a constant current or barrier distance, the record of the vertical tip motion will reflect the surface topography.

The success of this technique rapidly gave birth to a large family of instruments generally referred to as *scanning probe microscopes* (SPM). Each member of this family uses a different type of interaction between the probing tip and the sample. The most popular ones are the STM, the AFM, and the scanning near-field optical microscope (SNOM). The SPM family works on a principle similar to a record player. A sharp tip (e.g. silicon nitride in AFM, diamond in a record player) is traversed across the surface (the sample, or the record). The interaction of the tip with the surface is measured and converted into an electrical signal which is processed into interpretable results (three-dimensional image of sample topography, or music from stereo speakers). However, unlike the record player, the sensing tip of an AFM is raster across the sample (much like how a television image is produced) rather than following a predefined spiraling track. In addition to simple topographic imaging, many modern AFMs have the capability to image via frictional force, phase contrast, and elasticity. Electrostatic, magnetic, and thermal imaging can also be performed with the appropriate equipment.

Measuring the ultra-small forces on particles as small as single atoms was a big problem. Binnig and Rohrer [3] proposed to do this by monitoring the elastic deformation of various types of springs with the scanning tunneling microscope. It was a common practice to use the displacement of springs as a measure of force, and previous methods had relied on electrostatic fields, magnetic static fields, optical waves, and X-rays. Jones [4] reviewed devices that use variable capacitances and reported that displacements of 10^{-4} Å can be measured. Tabor and co-workers [5,6] used optical interference methods to measure displacement of 1 Å. Deslattes [7] measured displacement of 10^{-6} Å, which is about 1% of the nuclear diameter, with an X-ray interferometer constructed from a single crystal of silicon.

Binnig et al. [1] proposed using the scanning tunneling microscope as a method to measure forces as small as 10^{-18} N. On this concept they introduced a new type of microscope capable of investigating the surfaces of insulators on an atomic scale. Their preliminary results *in air* demonstrated a lateral resolution of 30 Å and a vertical resolution less than 1 Å. In their system, the STM was used to measure the motion of a cantilever beam with an ultra-small mass. The force required to move this beam through measurable distances (10^{-4} Å) could be as small as 10^{-18} N. This level of sensitivity clearly penetrates the regime of interatomic forces between single atoms and opens the door to a variety of applications. Thus, the AFM is a new tool designed to exploit this level of sensitivity.

On the basis of Binnig et al.'s [1] investigations, many types of AFM have been commercialized. A few companies who manufacture AFMs are listed in Table 3.1.

Table 3.1. List of a few companies that manufacture atomic force microscopes

	Company and address	Type of microscope
1	Veeco Instruments Inc. (Digital Instruments) 100 Sunnyside Blvd., Ste. B Woodbury, NY 11797-2902, USA	SPM, AFM, STM, and SFM
2	JPK Instruments AG Bouchéstrasse 12 Haus 2, Aufgang C 12435 Berlin, Germany	SPM
3	Nanograph Systems School of Physics & Astronomy, University Park, Nottingham NG7 2RD, UK	SPM
4	Nanonics Imaging, Ltd. Manhat Technology Park Malcha, Jerusalem, Israel, 91487	SPM, AFM
5	Novascan Technologies, Inc. 131 Main Street Ames, IA 50010, USA	AFM
6	Rastersonden und Sensormesstechnik GmbH (Surface Imaging Systems) Kaiserstrasse 100 (Technologiepark Herzogenrath), TPH D-52134 Herzogenrath, Germany	SPM
7	Nanotec Electronica Centro Empresarial Euronova 3 Ronda de Poniente, 2 Edificio 2 - 1ª Planta - Oficina A28760 Tres Cantos Madrid, Spain	SPM, STM, and AFM
8	WITec GmbH Hoervelsinger Weg 6 89081 Ulm, Germany	AFM and pulsed force mode AFM
9	Infinitesima Oxford Centre For Innovation Mill Street, Oxford OX2 0JX, UK	High-speed SPM
10	Molecular Imaging, Inc. 4666 S. Ash Avenue Tempe, AZ 85282, USA	AFM and force mode AFM
11	NT-MDT Co Technopark, Zelenograd Moscow, Russia	AFM and STM
12	Seiko Instruments, Japan	AFM

3.1.1 Terms and Abbreviations

The following are the terms and abbreviations which are widely used for atomic force microscopes:

Cantilever Flexible portion of the probe extending from the substrate and to which the tip is attached
DSP Digital signal processor. Computer processor used to control SPM feedback loop
EC-AFM Electrochemical atomic force microscopy
Feedback Process of self-correction between the probe's actual, real-time height–surface force and its intended height–surface force, based upon the probe's signal
Fluid cell Accessory used for imaging materials in fluid, consisting of a specialized tip holder and O-ring
LFM Lateral force microscopy (frictional measurements of surfaces based upon a tip's lateral and torsional response)
Probe Integrated mechanical device used to image surfaces; includes a substrate, cantilever, and tip
SPM Scanning probe microscopy (a general term encompassing all types of microscopy which utilize a scanned micro-sharpened probe and feedback circuitry to image nanometric phenomena, including AFM, ECAFM, LFM, STM, and many others)
Sensitivity Amount of movement produced by a piezo scanner for a given amount of voltage
Spring constant Amount of force required to bend a cantilever some given amount
Tip holder Removable appliance for mounting SPM probes (on AFMs, the tip holder is installed within the head of the microscope)

3.1.2 Advantages and Disadvantages of AFM

The following are the advantages of AFM:

1. It enables quantitative surface measurement.
2. It can image any solid surface without any special sample preparation.
3. It can measure physical forces.
4. Compared to scanning electron microscopy (SEM), it can provide more accurate topographic contrast, direct height measurements, and unobscured views of surface features (no coating necessary).
5. Compared with transmission electron microscopy (TEM), it can provide three-dimensional images without expensive sample preparation and yield far more complete information than the TEM profiles available from cross-sectioning samples.
6. Compared with optical interferometric microscopes (optical profiles), it may provide unambiguous measurement of step heights, independent of reflectivity differences between materials.

7. Quantitative topographic and surface property determinations provided by AFM may be correlated (via multivariate statistical procedures or neural network analyses) with other independent measurements of membrane surface properties, such as chemical or microbial adsorption data, water flux, solute transport, surface energy, etc.

The main disadvantage of AFM, compared to the electron microscope, is the image size. The electron microscope can show an area on the order of millimeters by millimeters and a depth of field on the order of millimeters. The AFM can only show a maximum height on the order of micrometers and a maximum area of around 125 × 125 μm. Other disadvantages include slow scanning, having to fix samples, and artifacts.

3.2 AFM: Principles and Applications

The AFM consists of a cantilever with a sharp tip at its end. The tip is brought into close proximity of a sample surface. The tip scans over the surface of the sample, its position and cantilever deflection are recorded, and a surface image is recorded.

3.2.1 AFM Principles

AFM images are obtained by measurement of the force on a sharp tip created by its proximity to the surface of the sample. This force is kept small and at a constant level with the feedback mechanism. When the tip is moved sideways, it follows the surface contours such as trace *B* in Fig. 3.1. Label '*A*' refers to an adsorbed site of a single atom in the gap of a scanning tunneling microscope.

The basic objective of the operation of the AFM is to measure the forces (at the atomic level) between a sharp probing tip and a sample surface (Fig. 3.1). Scanning the sample relative to the probing tip and measuring the deflection of the cantilever as a function of lateral position produces images. Typical spring constants (amount of force required to bend a cantilever some given amount) are between 0.001 and 100 N/m, and motions from microns to ~ 0.1 Å arc measured by the deflection sensor. Typical forces between the tip and the sample range from 10^{-11} to 10^{-6} N. For

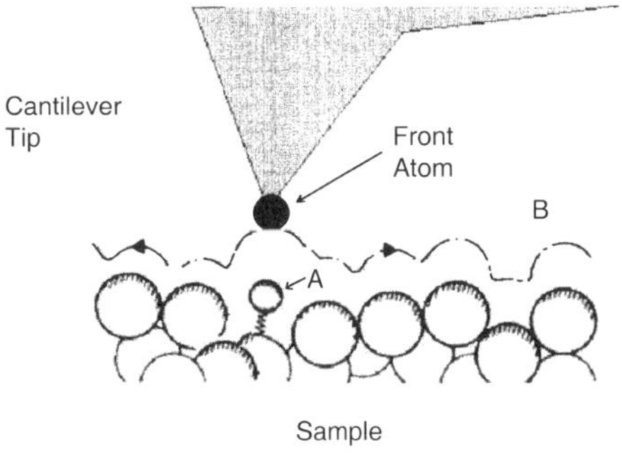

Fig. 3.1. Mechanism of AFM

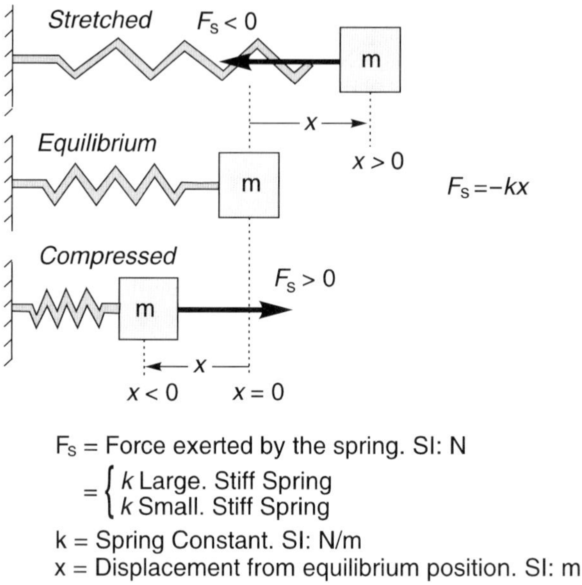

Fig. 3.2. Hooke's law—spring force

comparison, the interaction between two covalently bonded atoms is of the order of 10^{-9} N at separations of ∼ 1 Å. Therefore, non-destructive imaging is possible with these small forces [8]. It was suggested that the scanning tunnelling microscope could be used to measure forces as small as 10^{-18} N [1]. A flexible cantilever with a very low spring constant could be produced. With a cantilever that induces forces smaller than interatomic forces, the topography of the sample could be measured without replacing the atom. The force between the tip and the sample leads to a deflection of the cantilever according to Hooke's law (Fig. 3.2).

Hooke's Law

An object is connected to a spring whose spring constant k can be changed along with the object's initial position. Displayed in Fig. 3.2 is the spring's force on the object as well as the object's position and velocity as a function of time.

The negative sign indicates that the spring force is a restoring force, i.e., the force F_s always acts in the opposite direction from the direction in which the system is displaced. Here, we assume that the positive values of x are the same as the positive values of the force.

The origin has to be placed at the position where the spring system would be in static equilibrium for the equation $F_s = -kx$ to be valid. This is the location where the net force on the object to which the spring is attached is equal to zero. If not, then $F_s = -k(x - x_0)$ where x_0 is the equilibrium position relative to the origin.

Springs are normally assumed to be massless so their inertia can be neglected. This also means that the forces exerted by both ends of the spring are the same but in opposite directions.

Thus, AFM incorporates a number of refinements listed here that enable it to achieve atomic-scale resolution:

1. Sensitive detection
2. Flexible cantilevers
3. Sharp tips
4. High-resolution tip–sample positioning
5. Force feedback

Figure 3.3 shows the outlines of the optical sensing system for contact mode AFM and LFM.

If the tip were scanned at a constant height, there would be a risk that the tip would collide with the surface, causing damage. Hence, in most cases, a feedback mechanism is employed to adjust the tip to sample distance to keep the force between the tip and the sample constant. This can be achieved by mounting the sample on

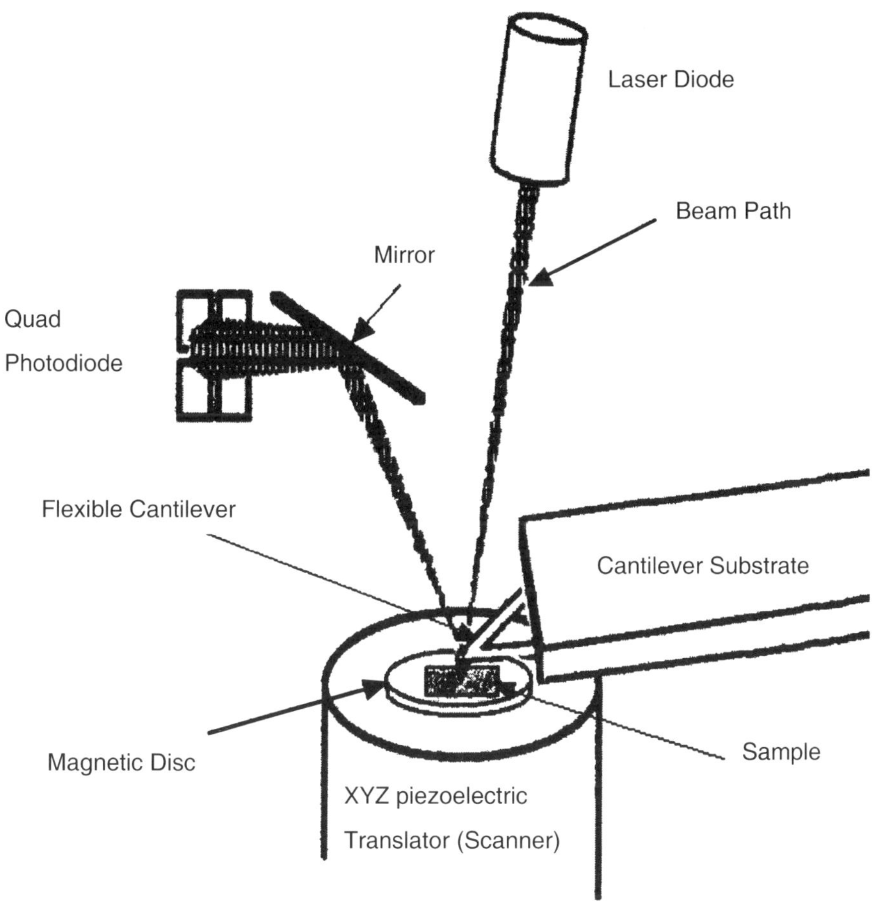

Fig. 3.3. Tapping mode and LFM optical sensing system

a piezoelectric crystal. The tip is then scanned across the sample surface, and the vertical displacement necessary to maintain a constant force on the tip is recorded. The resulting map $z(x, y)$ represents the topography of the sample.

During experiments, the cantilever gives constant deflection, and hence, the force applied by the stylus to the sample remains constant. This deflection off the cantilever is measured by detecting the angular deflection of a laser beam reflected off the back of the cantilever. A light beam is used from a laser diode. The reflection from the back of the cantilever is picked up by a quartered photodetector. The intensity on the different segments of the photodetector is used as a deflection signal. The choice of segments depends on the mode of AFM operation. Using this signal, feedback controls the z-motion of the piezoelectric scanner [9–13]. The images, which can include sample areas up to $125 \times 125\ \mu m$, can be stored in a computer and processed later.

3.2.2 Components of AFM Equipment

Figure 3.4 shows the AFM system hardware [9]. In AFM, there are seven major components:

1. Scanning probe microscope (SPM)
2. Controller
3. Computer
4. Keyboard
5. Mouse
6. Display monitor
7. Control monitor

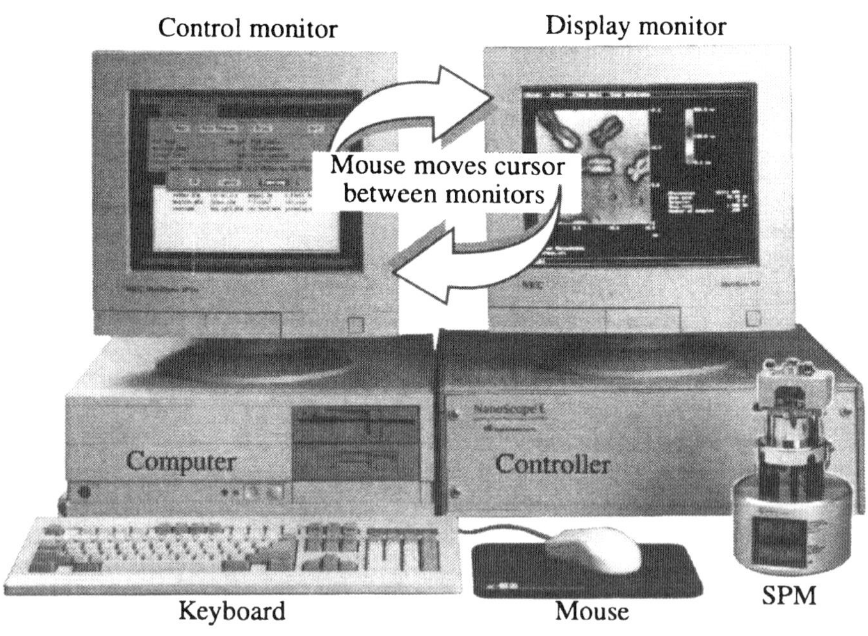

Fig. 3.4. MultiMode SPM system hardware. Reprinted from [9, 13]

Mouse movements automatically transfer the cursor between monitors, enabling the operator to seamlessly switch between control and display functions.

Figure 3.5 shows a schematic of the typical AFM tool (one of a few designs used). The major components are:

1. Thin cantilever with extremely sharp probing tip (10–50 Å in radius). The style and shape of the cantilever will vary depending on the operating mode.
2. Three-dimensional piezoelectric scanner.
3. Optical system to measure deflection of the cantilever.

Figure 3.6 shows a front view of the MultiMode microscope and its major components, developed by Digital Instruments, Inc. (Santa Barbara, California, USA) [9].

Looking at Fig. 3.6, the top square block is the SPM head that is detachable from the piezoelectric scanner. The details of the SPM head and the laser beam path are shown in Fig. 3.7. The figure also shows various adjustment knobs. The head and attached x, y-stage are kinematically mated to the scanner via three contact points. A pair of retaining springs holds down the head, allowing it to be raised and lowered using adjustment screws threaded through the scanner body.

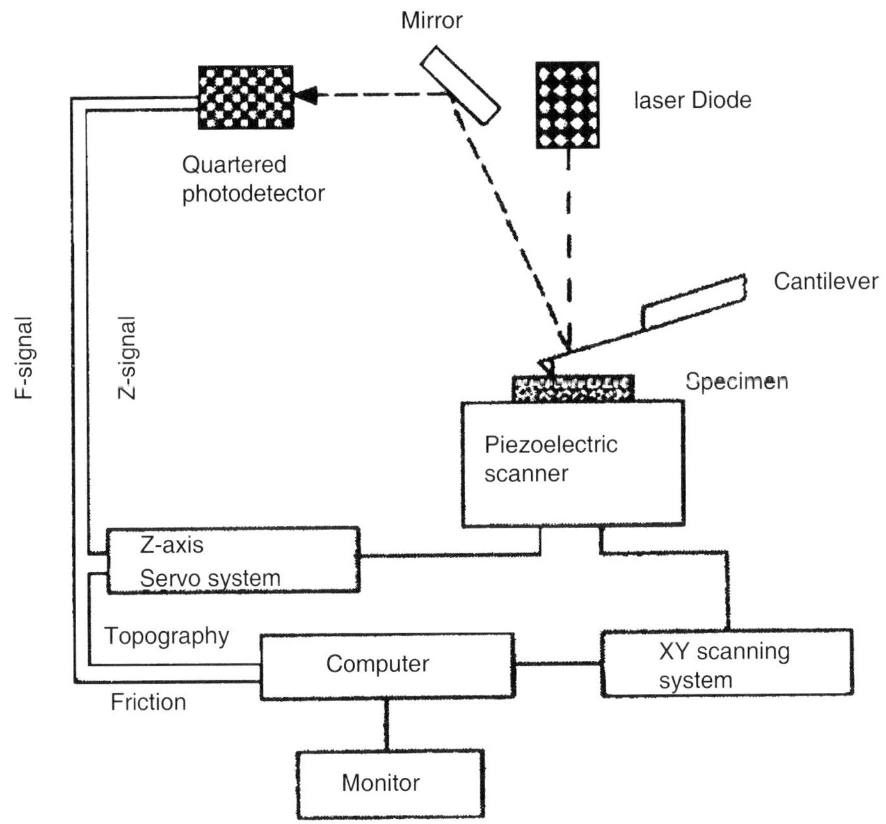

Fig. 3.5. AFM system scheme

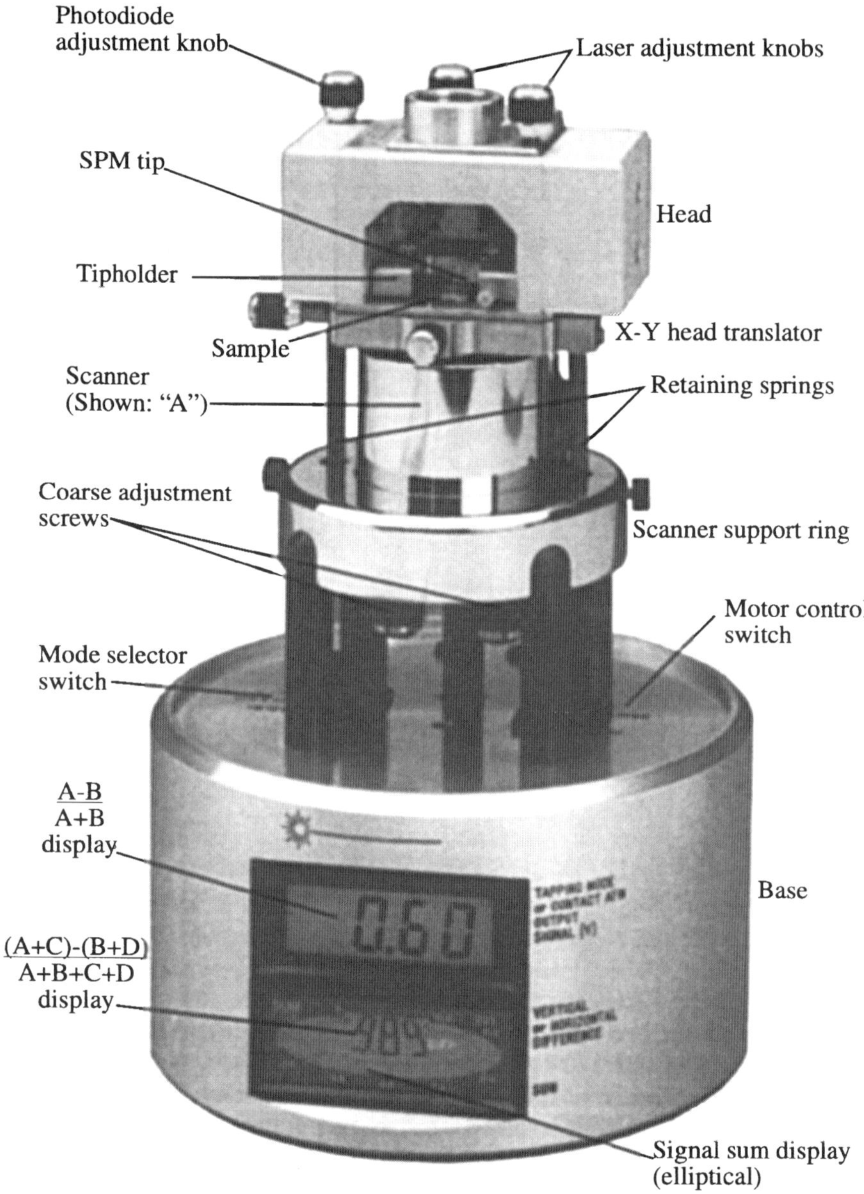

Fig. 3.6. MultiMode SPM. Reprinted from [9, 13]

Let us now follow the laser beam path. The beam from a laser diode (*1*) is focused onto the back of the cantilever (*3*) with the help of a mirror (*2*). The beam reflects off the back of the cantilever onto a segment photodiode (*5*) with the help of another mirror (*4*). The amplified differential signal between the upper and lower photodiodes provides a sensitive measure of the cantilever deflection.

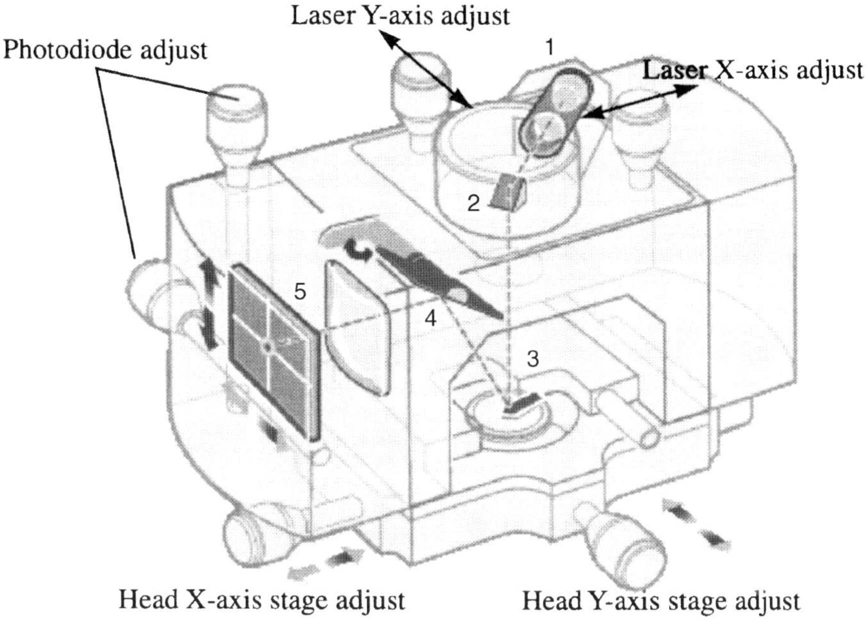

Fig. 3.7. MultiMode SPM head and major components: laser (*1*); mirror (*2*); cantilever (*3*); tilt mirror (*4*); and photodetector (*5*). Reprinted from [9, 13]

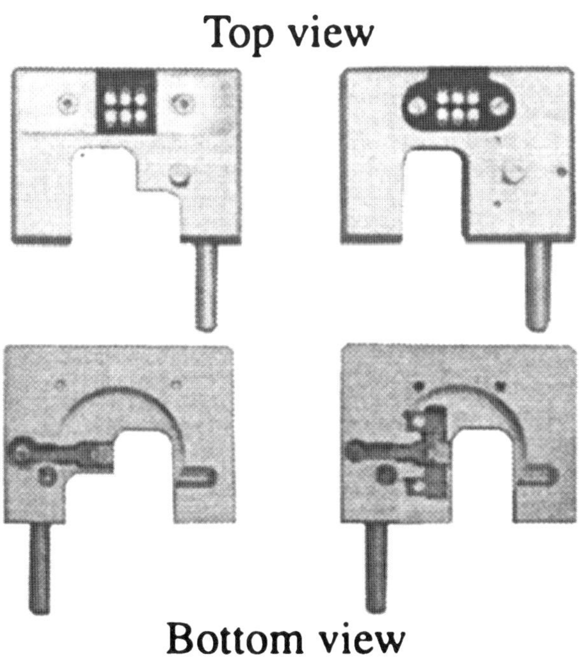

Fig. 3.8. Cantilever holders. Reprinted from [9, 13]

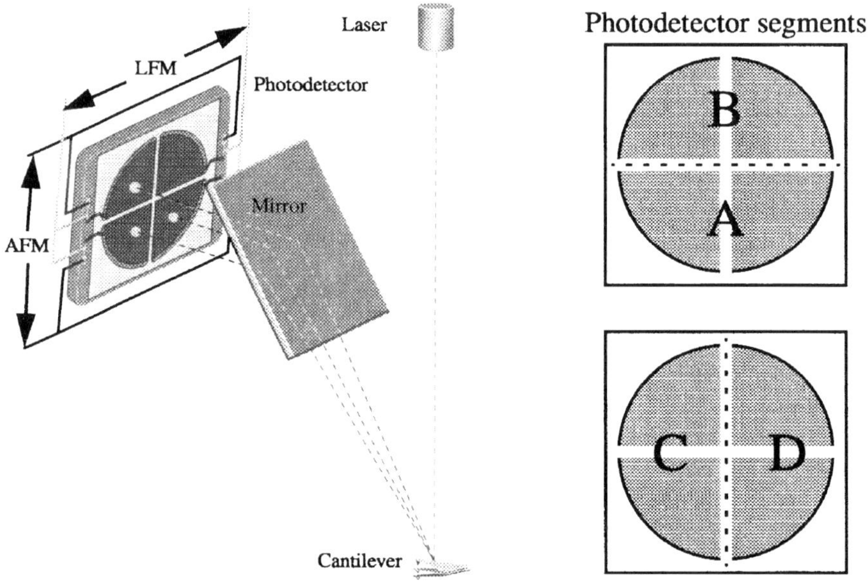

Fig. 3.9. Quad photodetector arrangement. Different segments of the photodetector are used for generating AFM and LFM signals. Reprinted from [9, 13]

The cantilever should be held in a tip holder. The top and bottom views of two tip holders are shown in Fig. 3.8. Cantilevers come in a variety of sizes, shapes, and materials and are chosen according to the type of imaging to be done.

Figure 3.9 shows the quad photodetector [9]. The photodiode is divided into four segments, all of which are combined to provide different information depending on the operating mode. In all modes, the four elements combine to form the sum signal. The amplified differential signal between the top two elements and the two bottom elements provides a measure of the deflection of the cantilever. This differential signal is used directly in contact AFM.

Looking at Fig. 3.6, you will find a scanner below the SPM head. AFM has various interchangeable scanners. The maximum scan size and resolution of images depend upon the choice of scanner. Each scanner exhibits its own unique piezo properties; each has its own parameter file. When scanners are changed, the parameter file for the new scanner is changed along with it, ensuring maximum accuracy at any scan size. Below the piezoelectric scanner is the instrument base with step motor, amplifiers, mode switch, and displays.

3.2.3 Different AFM Modes

3.2.3.1 Forces Working in AFM

In AFM, several forces contribute to the deflection of the cantilever. The force most commonly associated with AFM is an interatomic force called the van der Waals forces. Figure 3.10 shows the dependence of the short-range repulsive force and the long-range van der Waals forces on the distance between the tip and the sample.

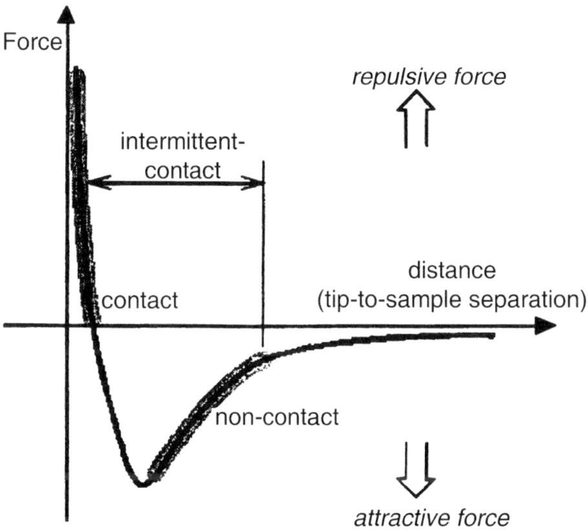

Fig. 3.10. Interatomic forces vs. distance curve

Two distance regimes are labeled on Fig. 3.10: the contact regime and the non-contact regime. In the contact regime, the cantilever is held less than a few angstroms from the sample surface, and the interatomic force between the cantilever and the sample is repulsive. In the non-contact regime, the cantilever is held on the order of tens to hundreds of angstroms from the sample surface, and the interatomic forces between the cantilever and the sample is attractive (largely a result of the long-range van der Waals interactions).

Laser beam deflection offers a convenient and sensitive method of measuring cantilever deflection. In the non-contact mode, the AFM derives topographic images from measurements of attractive forces; the tip does not touch the sample. On the other hand, in the contact mode, repulsion forces between the tip and the sample produce topographic images.

3.2.3.2 AFM Modes of Operation

The AFM can be operated in many ways. The main classes of interaction are *contact mode*, *tapping mode*, and *non-contact mode*. Table 3.2 shows the modes of operation for AFM and the types of forces of interaction working in the individual modes of operation.

Table 3.2. Mode of operation for AFM and the forces of interaction working in each mode

Mode of operation	Force of interaction
Contact mode (C-AFM)	Strong (repulsive)—constant force or constant height
Non-contact mode (NC-AFM)	Weak (attractive)—vibrating probe
Intermittent contact mode (TM-AFM)	Strong (repulsive)—vibrating probe
Lateral force mode	Frictional forces that exert a torque on the scanning cantilever
Magnetic force	Magnetic field of the surface
Thermal scanning	Distribution of thermal conductivity

For membranes, mainly contact mode (C-AFM), non-contact mode (NC-AFM), and tapping mode (TM-AFM, intermittent contact mode) are used.

3.2.3.3 Contact Mode

In C-AFM, the tip makes physical contact with the sample. As the tip is moved across the sample, the contact force causes the cantilever to bend according to changes in topography. In constant force mode, the tip is constantly adjusted to maintain a constant deflection, and therefore constant height above the surface. It is this adjustment that is displayed as data. However, the ability to track the surface in this manner is limited by the feedback circuit. Sometimes the tip is allowed to scan without this adjustment, and one measures only the deflection. This is useful for small, high-speed atomic resolution scans, and is known as *variable-deflection* mode.

The lateral forces acting between the tip and the sample in contact mode are used to examine the friction of relatively flat surfaces, whereas the lateral force images of corrugated samples can help to visualize morphological features (lateral force microscopy). The lateral forces, however, increase the mechanical surface damage. In conventional C-AFM, the probe tip is simply dragged across the surface of the sample. It has some serious drawbacks. The dragging motion of the probe tip, combined with adhesive forces between the tip and the surface, can cause substantial damage to both the sample and the probe and create artifact data.

In general, C-AFM:

1. Provides three-dimensional information nondestructively, with 1.5 nm resolution laterally and 0.05 nm resolution vertically
2. Uses strong repulsive forces acting between the tip and the sample
3. Analyzes insulators and conductors easily (AFM is not based on conductivity)
4. Operates in air and fluid environments
5. Provides information about physical properties—elasticity, adhesion, hardness, friction, etc.

3.2.3.4 Non-contact Mode

NC-AFM belongs to a family of AC modes, which refers to the use of an oscillating cantilever. A stiff cantilever is oscillated in the effective regime, meaning that the tip is quite close to the sample, but not touching it (*non-contact*).

In NC-AFM, the stiff cantilever oscillates near the surface of the sample at a frequency of 50 000–500 000 cps. The tip has no contact with the sample. The cantilever is held 5–10 nm away from the surface, within the region of the force distance curve where the long-range van der Waals forces are dominant [7]. In this mode of operation, the tip is responding to a force between the tip and the sample and can be several orders of magnitude lower than the force in contact mode.

The non-contact mode AFM was developed by Martin et al. [14]. It profiles a surface in a different fashion than the contact AFM. NC-AFM is desirable in studying the membrane surface, because synthetic membranes are mostly made of polymers, which make the surface soft [15, 16]. Stiff cantilevers are used in NC-AFM studies because the soft cantilevers can be pulled into contact with the sample surface. However, the use of stiffer cantilevers reduces the change in cantilever deflection and vibrational amplitude, and thus, a sensitive detection scheme is needed. It should be emphasized that the use of non-contact AFM can allow imaging of membrane surfaces that cannot be imaged in contact mode.

3.2.3.5 Tapping Mode

Tapping mode is also commonly referred to as intermittent contact or dynamic force mode (DFM). The need to avoid surface damage was one of the major motivations leading to development of TM-AFM, first introduced by Zhong et al. [17]. In this mode, a stiff cantilever is oscillated closer to the surface than in non-contact mode. Part of the oscillation extends into the repulsive regime, so the tip intermittently touches or taps the surface. Very stiff cantilevers are typically used, as tips can get stuck in the water contamination layer. The advantage of tapping the surface is an improved lateral resolution on soft samples. Lateral forces such as drag, common in contact mode, are virtually eliminated. This technique is less likely to damage the membrane surface than C-AFM. It is more effective than NC-AFM for imaging larger scan sizes that may include larger variations in topography.

In TM-AFM, amplitude damping of a fast-oscillating probe is employed for surface imaging, and a short, intermittent tip–sample contact prevents development of inelastic surface deformation. Operation of tapping mode under water and use of small oscillation amplitudes are ideal for successful imaging of soft sample. In tapping mode, a cantilever oscillates in free air at its resonant frequency. A piezo stack excites the cantilever's substrate vertically, causing the tip to bounce up and down. As the cantilever bounces vertically, the reflected laser beam is deflected in a regular pattern over a photodiode array, generating a sinusoidal electronic signal. The signal is converted to a root mean square (rms, R_q) amplitude value, which is displaced in AC volts. Figure 3.11a represents a cantilever oscillating in free air at its resonant frequency. Figure 3.11b represents the same cantilever at the sample surface. Although the piezo stack continues to excite the cantilever's substrate with the same energy, the tip is deflected in its counter with the surface. The reflected beam (return signal) reveals information about the vertical height of the sample surface and some characteristics of the sample material itself. These material characteristics may include elasticity (hardness) and the magnetic and/or electric force present. AFM can image surfaces in air and under liquids without special surface preparation. Resolution can reach atomic dimensions for flat surfaces [13].

Tapping mode imaging is an advance in AFM of soft, adhesive, or fragile samples. Digital Instruments, Santa Barbara, California, developed this technique. It allows high-resolution topographic imaging of sample surfaces that are easily damaged, loosely held to their substrate, or otherwise difficult to image by other AFM techniques. Tapping mode overcomes problems associated with friction, adhesion, electrostatic forces, and other difficulties that can plague the conventional AFM scanning method. Tapping mode avoids the force instabilities caused by thermal drift in contact mode, resulting in time savings and improved image and measurement quality.

Tapping mode in fluids was first introduced by the Hansma Research Group [18]. In the first implementation of tapping mode in fluids, the sample, which sits on a piezoelectric scanner, oscillates up and down and taps the tip at the apex of each

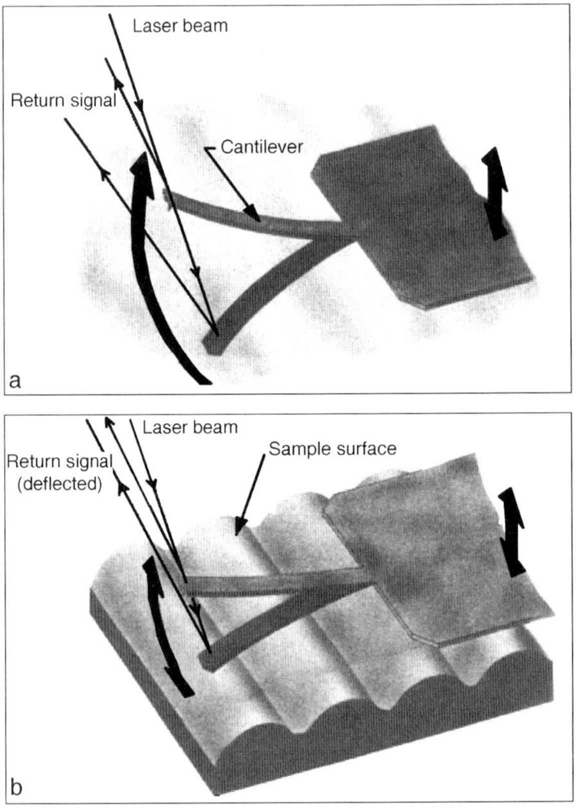

Fig. 3.11. **a** Tapping cantilever in free air. **b** Tapping cantilever on sample surface. Note: deflection of cantilever and return signal are exaggerated. Reprinted from [9, 13]

oscillation cycle. The amplitude of the piezoelectric scanner is set manually at the beginning of the run, and the tapping force is held constant by a feedback loop. In general, TM-AFM:

1. Measures composition, adhesion, friction, and viscoelastic properties by phase lag
2. Identifies two-phase structure of polymer blends
3. Is less damaging to soft samples than lateral force microscopy
4. Identifies surface contaminants that are not seen in height images

It should be noted that for high-resolution imaging and most routine topographic profiling, the systems are kept in direct contact with the surface. The non-contact or tapping mode method has been used to image magnetic and electronic fields, liquid films, and soft surfaces (for example polymeric membranes).

3.2.4 More Information about the Cantilever

Cantilevered probes are the most important component of the scanning probe microscope. Hence, more information is provided for the cantilevered probes. These con-

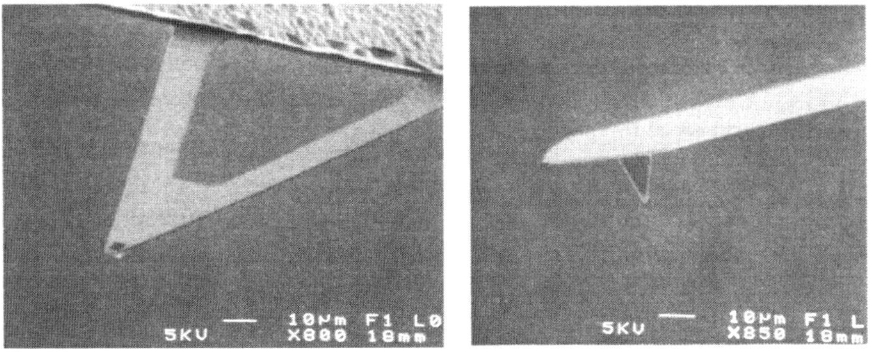

Fig. 3.12. Two types of cantilevered probes: silicon nitride, *left*, and crystal silicon, *right*. Reprinted from [9, 13]

sist of a flexible cantilever from a rigid substrate, to which a tip has been attached. In AFM, the cantilever's flexibility acts as a nanometric spring, allowing the tip to measure surface forces.

Figure 3.12 shows two types of cantilevered probes: silicon nitride (*left*) and crystal silicon (*right*). Etched silicon probes are the most commonly used probes for TM-AFM applications [9].

One of the most important factors influencing the resolution that may be achieved with an AFM is the sharpness of the scanning tip. The first tips used by the inventors of the AFM were made of diamond glued onto pieces of aluminum foil. Commercially fabricated probes are now universally used. The best tips may have a radius of curvature of only around 5 nm. The need of sharp tips is normally explained in terms of *tip convolution*. This term is often used (slightly incorrectly) to group together any influence which the tip has on the image. The main influences of the tip on the image are:

1. Broadening
2. Compression
3. Interaction forces
4. Aspect ratio

Tip broadening arises when the radius of curvature of the tip is comparable with, or greater than, the size of the feature trying to be imaged. Figure 3.13 illustrates this problem: as the tip scans over the specimen, the sides of the tip make contact with the feature. This is what we call tip convolution.

Compression occurs when the tip is over the feature trying to be imaged. It is difficult to determine in many cases how important this effect is, but studies on some soft biological polymers (such as DNA) have shown the apparent DNA width to be a function of imaging force. It should be kept in mind that although the force between the tip and the sample may only be nN, the *pressure* may be MPa.

Interaction forces between the tip and the sample produce the image contrast for the AFM. However, some changes which may be perceived as being topographical

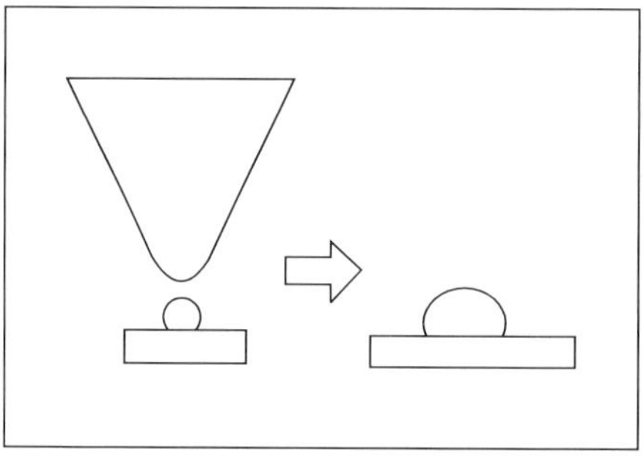

Fig. 3.13. Influence of tip-broadening

may be due to a change in force interaction. Forces due to the chemical nature of the tip are most important here, and selection of a particular tip for its material can be important.

The aspect ratio (or cone angle) of a particular tip is crucial when imaging steep sloped features. Electron-beam-deposited tips have been used to image steep-walled features far more faithfully than can be achieved with the common pyramidal tips.

Selection of soft cantilevers is a necessary step for AFM imaging, depending on the type of sample. The cantilever typically used in AFM is made of Si or Si_3N_4 and has integrated tips. With the aim of modulating several relevant properties, such as resonance frequencies or spring constants, the cantilevers can be prepared with different lengths, thicknesses, and shapes. As an example, cantilevers with small spring constants (so that tiny forces are able to produce large, detectable deflections) and high resonance frequencies (to avoid vibrational instabilities) are required for C-AFM [19]. The former is achieved by making the cantilever thin, whereas the latter is achieved by making the cantilever short. For TM-AFM under ambient conditions, stiffer cantilevers (several tens of Nm^{-1}) are needed to prevent the tip from getting stuck to the surface, mainly as a consequence of capillary forces [17]. The Si_3N_4 cantilevers with tips of pyramidal shape normally employed in C-AFM are suitable for the study of flat samples at the atomic as well as the micrometric scale [19].

Most users purchase AFM cantilevers with their attached tips from commercial vendors, who manufacture the tips with a variety of microlithography techniques. A close inspection of any AFM tip reveals that it is rounded off. Therefore, AFM microscopists generally evaluate tips by determining their *end radius*. In combination with tip-sample interaction effects, this end radius generally limits the resolution of AFM. As such, the development of sharper tips is currently a major concern.

The sharpened Si_3N_4 probes, which also have small spring constants ($\sim 0.04\ Nm^{-1}$), are a less expensive alternative to the Si probes. The size of the tip sample contact region depends also on the tip radius, which can be estimated by imaging standards [20, 21]. Accumulated knowledge of tip-sample force interactions has led to a better understanding of AFM use for polymer surfaces. The potential of AFM has been increased further by the recent development of a new imaging mode. Each of the TM-AFM probes consists of a short, stiff silicon cantilever with an integrated single crystal silicon tip. The cantilever has high resonance frequencies and

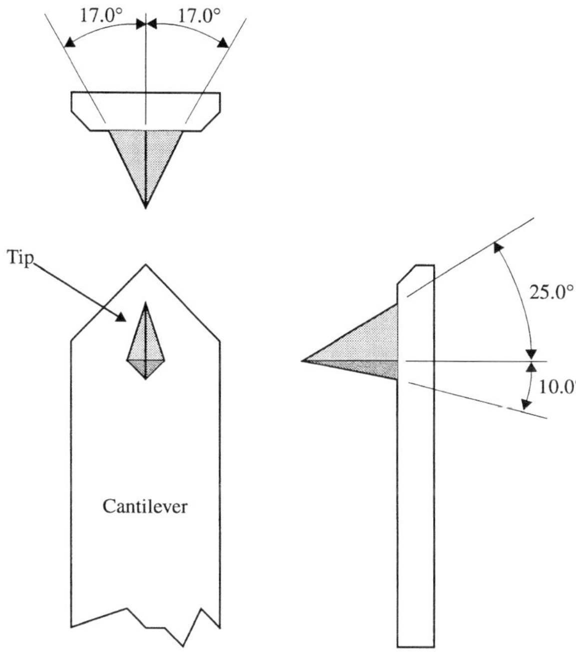

Fig. 3.14. Tapping mode etched silicon probe

Table 3.3. Tapping mode etched silicon probe specifications

Force (or spring) constants	20–100 Nm^{-1}
Resonant frequency	200–400 kHz
Normal tip radius of curvature	5–10 nm
Cantilever length	125 μm
Cantilever configuration	Single beam
Reflective coating	Uncoated
Tip half angle	17° side, 25° front, and 10° back

high spring constants. The geometry of an etched Si tip is given in Fig. 3.14. Details are given in Table 3.3 [22].

Etched silicon probes provide the highest aspect ratio and the most consistent tip sharpness of the probes supplied at present. The silicon nitride probe is inexpensive, durable, and suitable for contact mode imaging.

The silicon nitride tip is used mostly for C-AFM. Measurements can be done in ambient air, controlled atmospheres, or in non-aggressive liquids. AFM also allows surface forces, and even molecular forces, to be directly quantified [23]. For example, the interaction forces between a silicon tip and microfiltration and ultrafiltration membranes in an electrolyte solution can be measured [24]. The geometry of the cantilever is not simple, and in some cases not even known, so comparison with theory is difficult. However, attaching a sphere to the cantilever instead of a tip enables the measurement of interaction between surfaces of known geometry [25]. This technique has been used to measure interactions between different materials in air and solutions—for example, long-range electrical double layer and London-van der Waals forces [23, 25–28].

3.2.5 Phase Imaging and Roughness Parameters

3.2.5.1 Image Display by AFM

AFM gives a three-dimensional image from the height image data. The usual method for displaying the data is to use a color mapping for height—for example, black for low features and white for high features. Similar color mappings can be used for non-topographical information such as phase or potential.

3.2.5.2 AFM Imaging

In AFM imaging, tip–sample interactions essentially can be modified by surface forces. This can help to reveal the spatial distribution of different component systems such as polymer blends and composites. For example, correlations were found between surface chemical structure, hydrophilicity/hydrophobicity, adhesion, and lateral forces [29,30]. If one takes into account differences in chemical structure, as well as possible variations in local hardness of the hydrophilic and hydrophobic domains, then image contrast may be correlated with functional properties and the chemical nature of the polymer surface. By using a chemically modified tip, the contribution of specific tip–sample force interactions in the image contrast can be enhanced. However, the differences in the image contrast can also originate from variations in molecular packing in the chemically homogeneous sample.

3.2.5.3 Phase Imaging

More recently, there has been much interest in *phase* imaging. This works by measuring the phase difference between the oscillations of the cantilever driving piezo and the detected oscillation. It is thought the image contrast is derived from image properties such as stiffness and viscoelasticity.

3.2.5.4 Roughness Parameters

Surfaces can be compared in terms of the roughness parameters, such as the mean roughness R_a, the mean square of the Z data R_q, and the mean difference in height between the highest peaks and five lowest valleys R_z, as well as in terms of the diameter of the nodules. The Z is defined as the difference between the highest and lowest points within the given area. The roughness parameters depend on the curvature and the size of the TM-AFM tip, as well as on the treatment of the captured surface data (plane fitting, flattering, filtering, etc.). Therefore, the roughness parameters should not be considered as absolute roughness values.

The mean roughness is the mean value of the surface relative to the center plane, the plane for which the volumes enclosed by the image above and below this plane are equal, and is calculated as

$$R_a = \frac{1}{L_x L_y} \int_0^{L_x} \int_0^{L_y} |f(x,y)| \, dx \, dy \tag{3.1}$$

where $f(x, y)$ is the surface relative to the center plane, and L_x and L_y are the dimensions of the surface.

The root mean square of the Z values R_q is the standard deviation of the Z values within the given area and is calculated as

$$R_q = \sqrt{\frac{\Sigma(Z_i - Z_{avg})^2}{N}} \tag{3.2}$$

where Z_{avg} is the average of the Z values within the given area, Z_i is the current value, and N is the number of points within a given area.

3.2.5.5 Key Measurements from AFM

1. True three-dimensional surface topographic imaging
2. Complete image analysis of all surface features or irregularities
3. Surface elasticity or compressibility measurements
4. Surface adhesion measurements
5. Quantitative summary statistics

3.3 Instructions for AFM Experiments

Generally, instructions for AFM experiments are provided by vendors with AFM manuals. One such set of instructions is summarized below [9,13].

Prepare the cantilever Etched silicon cantilever substrates are generally used for NC-AFM or TM-AFM, and silicon nitride cantilevers are used for C-AFM. In both cases, the cantilever probe should be inspected under the microscope when being used for first time. Use the sharp-pointed tweezers to remove the cantilever substrate from the container. Grasp the sides of the substrate, away from the lever and probe tip. Be very careful about avoiding any contact with the probe lever, since it will immediately snap off. Silicon is very brittle.

Prepare the sample The calibration sample or other sample should be placed on one of the 15-mm-diameter metal disks used for sample mounting. Before putting the sample on the metal disk, put double-sided adhesive on the disk. Using tweezers, place a small sample to be imaged firmly on the "stickytab" adhesive and gently press until the sample is secured. Alternatively, a small sample can be glued down to the sample puck using cyanoacrylate glue (superglue). Place the small sample disk atop the scanner.

Load the sample Remove the head of the AFM by unfastening the retaining springs on either side and unplugging the head's connector. Lift the head off and set it aside. This will expose the top of the scanner tube. Mount the sample puck on the scanner tube. An internal magnet supplied with most units holds the puck down. With the sample in place, remount the head by gently lowering it over the scanner tube. The top of the sample should protrude no more than a few millimeters above the top of the head's x, y translation stage. Secure both retaining springs and plug the head's connector into the support ring. Check the head for free vertical movement (Fig. 3.15).

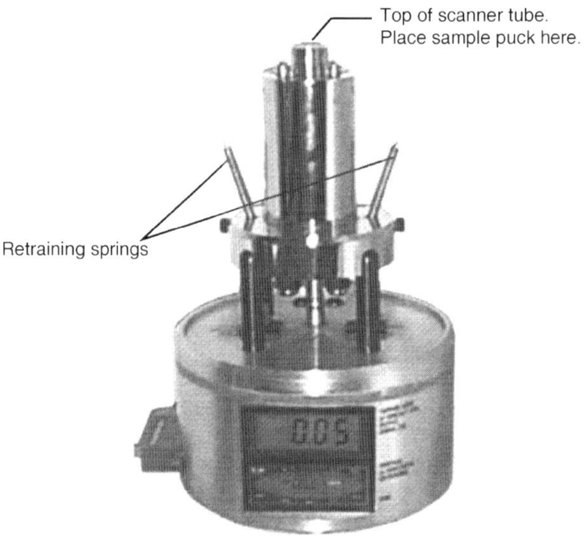

Fig. 3.15. MultiMode base with scanner mounted on support ring. Reprinted from [9, 13]

Load the probe into the tip holder Details on how to load the silicon nitride probe into the tip holder are given in the instruction manual [9]. The procedure for installation of a single crystal silicon probe for TM-AFM is essentially the same as for the installation of a silicon nitride probe for C-AFM. The substrate should be face up, with the probe's cantilever pointing away from the AFM tip holder. This ensures that the cantilever and tip are facing toward the sample once the tip holder is mounted in the head. Slide the probe with the help of tweezers into the tip holder's groove. Gentle downward pressure against the tip holder will lift the spring clip for probe insertion

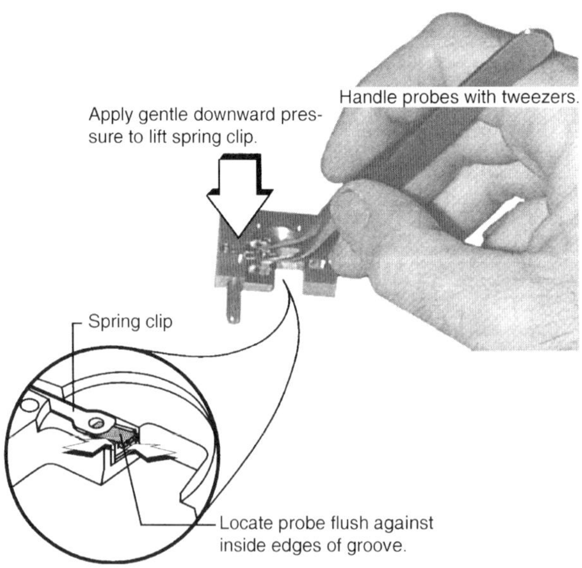

Fig. 3.16. Underside of AFM tip holder. Slide the probe carefully into the tip holder's groove. Gentle downward pressure against the tip holder will lift the spring clip for probe insertion. Reprinted from [9, 13]

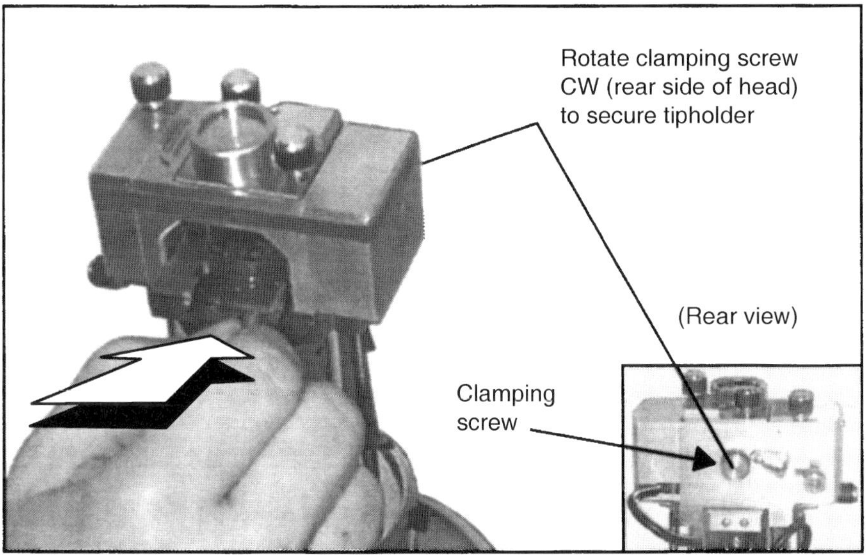

Fig. 3.17. Install tip holder in head without touching the sample. Secure tip holder using clamp screw at rear of head. Reprinted from [9, 13]

(Fig. 3.16). Fluid cell probe installation is similar to AFM tip holders (details are in the manual). Install the tip holder in the head without touching the sample. Secure the tip holder using the clamping screw at the rear of the head (Fig. 3.17) [9].

Laser alignment There are two methods for aligning the laser, mirror, and photodiode for all modes except STM, since STM does not use a laser. The first method uses a high-powered monocular magnifier to observe the laser spot's position on the cantilever. The second method uses a strip of paper to observe the laser's position. The choice of method is largely a matter of personal preference.

(1) The magnifier method In this method, a high-powered monocular magnifier or a similar magnifying system is used. First, the laser spot is positioned onto the cantilever. The photodiode is then positioned to maximize the signal, and the spot is fine adjusted onto the very tip of the cantilever (Fig. 3.18) [9,13].

(2) The paper method Most users prefer to align the laser by observing light patterns reflected or diffracted from the surface of the cantilever onto a piece of paper. This is known as the paper method of laser alignment. Interpreting light patterns requires some experience, but it is not difficult. The paper method can quickly verify with a high level of confidence whether the laser spot is on the tip of the cantilever.

The general method is as follows:

1. Locate the laser spot on the substrate.
2. Move the spot along the y-axis to locate the center of the cantilever.
3. Move the spot off the substrate and onto the cantilever.
4. Locate the spot on the center of the cantilever.

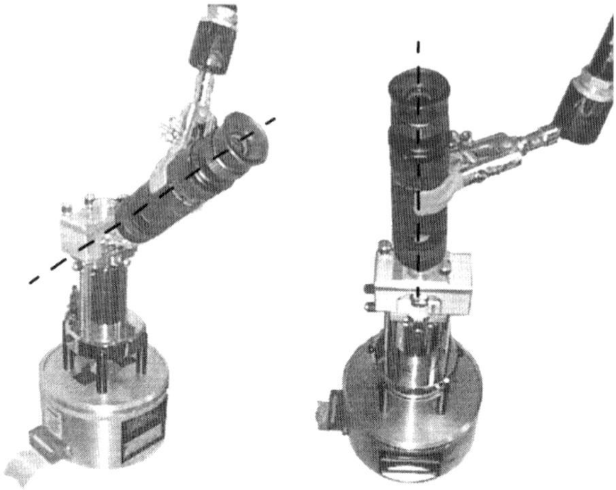

Fig. 3.18. Two ways of positioning a magnifier when aligning the laser: through the head's front windows, *left*, and overhead, *right*. Reprinted from [9, 13]

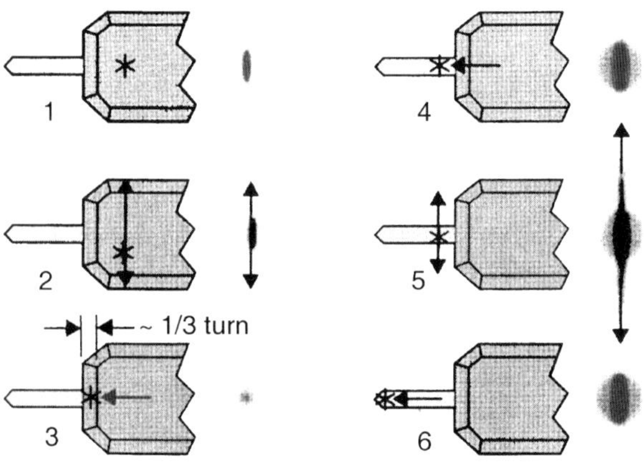

Fig. 3.19. Procedure for aligning silicon cantilevers using a slip of paper: **1** locate beam on substrate, **2** find center of substrate, **3** move toward cantilever, **4** locate cantilever, **5** find center of cantilever, and **6** move to end of cantilever. Reprinted from [9, 13]

5. Move the laser spot to the end of the cantilever.
6. Align the photodiode with the reflected beam.

Figure 3.19 can be used as a guide [13].

The manufacturer supplies details of both methods for the individual AFM machines. To avoid eye damage due to high-level laser light, in the first method, use a magnifier with a laser filter, and in the second, do not place highly reflective or metallized objects into the head area while the laser is on. *Avoid direct contact of laser beam with eyes.*

Photodiode Alignment After the laser beam is on the tip of the cantilever, adjust the photodiode positioner to maximize the sum signal on the elliptical bar graph (at the bottom of the scanner). This adjustment is much less sensitive than the laser position adjustment. The maximized value should be approximately 5 – 9 V for silicon nitride cantilevers. The value of this signal varies with many factors. It is important to note that it is possible to see a large response on the elliptical bar graph without having the laser beam on the cantilever. So it is important to visually verify that the laser beam is on the cantilever not relying on the elliptical bar alone. Attempting to engage with the laser beam improperly aligned will usually destroy the cantilever. Final adjustment to get a maximum sum signal can be done with the help of the photodiode and laser beam alignment knobs. The manufacturer provides the details for the adjustment of photodiode alignment.

3.4 AFM Applications for Synthetic Membranes

Atomic force microscopy was first applied to polymer surfaces in 1988, shortly after its invention [31]. Today, these studies range from relatively simple visualization of morphology to more advanced examination of polymer structures and properties on the nanometer scale.

The microscale surface features of polymer membranes influence colloidal and biofouling kinetics. AFM provides essential information about the submicron surface topography and fundamental material properties of commercial or experimental membranes. Such information has been correlated with the performance (flux and solute rejection) of RO/UF/MF membranes, permeation and selectivity of gas separation membranes, and fouling potentials of membranes. Such information is, therefore, critical in optimizing the functions of membranes and designing novel antifouling surfaces. The AFM is an excellent tool for examining the topography of polymer membrane surfaces in air-dried as well as fully hydrated form (under water also). AFM provides quantitative, three-dimensional images and surface measurements with a spatial resolution of a few micrometers down to a few angstroms.

NC-AFM is better than C-AFM for imaging small pores such as those in ultrafiltration and nanofiltration membranes. The reason for this is that the diameter of the cantilever tip apex is greater than the pore diameter. When the tip is passed over the small pore, the tip cannot penetrate into the pore, and there will not be a great change in cantilever deflection. However, TM-AFM is more successful at measuring the pore size and nodule size on the membrane surface.

The depressions in the AFM images of the membranes are considered to be pores; in gas separation membranes, they are called internodular domains, since there are no pores in the ordinary sense in those membranes. The mean size of the internodular domains is calculated by measuring the distance between two nodules present in the AFM image. Surfaces of membranes can also be compared in terms of the roughness parameter [17, 32].

3.5 Summary

AFM is based on the interaction forces (short- or long-range, attractive or repulsive) that exist between atoms and molecules, and these forces are present on all materials. AFM is optimized for measuring surface features that are extremely small, thus it is important to be familiar with the dimensions of the features being measured. AFM is capable of imaging features as small as a carbon atom and as large as the cross-section of a human hair. A carbon atom is approximately 0.25 nm in diameter, and a human hair is approximately 80 μm in diameter.

In principle, AFM resembles the record player as well as the stylus profilometer. It uses a very sharp pointed mechanical probe to collect real-space morphological information of solid surfaces. The tip is brought into close proximity of a sample surface. The force between the tip and the sample leads to a deflection of the cantilever according to Hooke's law. Typically, the deflection is measured using a laser spot reflected from the top of the cantilever. If the tip were scanned at constant height, there would be a risk that the tip would collide with the surface, causing damage. Hence, in most cases, a feedback mechanism is employed to adjust the tip-to-sample distance to keep the force between the tip and the sample constant. This can be achieved by mounting the sample on a piezoelectric crystal. The tip is then scanned across the sample surface, and the vertical displacement necessary to maintain a constant force on the tip is recorded. The resulting map of $z(x, y)$ represents the topography of the sample.

AFM images show critical information about surface features with unprecedented clarity. Atomic force microscopy is an instrument used for studying surface properties of materials at the atomic to micron level. It is attracting a great deal of interest because of its versatility and performance in a wide range of measurement and imaging applications. AFM is used in the electronic, telecommunications, biological, chemical, automotive, aerospace, medical, membrane, and energy industries. AFM can be used to investigate a variety of materials that include thick and thin film coatings, ceramics, composites, glass, synthetic and biological membranes, metals, polymers, and semiconductors. AFM may be used to image surfaces at atomic resolution as well as to measure forces at the nanonewton scale. From AFM, phase imaging goes beyond simple topographical mapping to detect variations in composition, adhesion, friction, viscoelasticity, pore size on the membrane surface, pore size distribution, and perhaps other properties. Applications include identification of contaminants and mapping of different components in composite materials and regions of low and higher surface adhesion or hardness. AFM gives three-dimensional images of the membrane surface including other properties. Contact mode AFM is good for mechanically stiff samples, or samples under fluid. Tapping mode AFM is good for a wider range of samples where either the sample is mechanically unstable or interaction with the tip produces poor results in contact mode. TM-AFM, at least with existing cantilevers, cannot be done under fluid.

The AFM cantilever is so thin and sensitive that it can sense the minute surface forces, such as van der Waals forces, magnetic forces, electrostatic forces, etc. It allows AFM to be used not only to investigate surface topography, but also to probe the physical, chemical, and magnetic properties of surfaces.

AFM is not limited to only conductive surfaces like STM. AFM is extremely flexible. It allows visualization of conductive, nonconductive, or semiconductive materials, and even living cells under a variety of environments (air, aqueous, and even corrosive conditions). In addition, AFM is capable of spatial resolution sufficient to visualize individual atoms at its smallest range (~Å) and is only limited by the scanning stage at its largest range (typically ~ 100 μm). AFM visualization requires neither special sample preparation nor expensive vacuum equipment (unlike STM). Finally, the instrument is quite compact, easily fitting within two cubic feet.

AFM also has limitations. In order to achieve a resolution on the order of angstroms, AFM needs substantial vibrational insulation, including both isolation tables and foam shielding to dampen air currents and sound waves, such as those produced by speaking humans. Spatial resolution in the z-axis is highly dependent on tip geometry. For a rough sample, a sharper tip is able to resolve smaller objects. Phase imaging can be invaluable for the mapping of surface hardness or elastic modules.

References

1. Bining G, Quate CF, Gerber C (1986) Phys Rev Lett 56:930
2. Bowen WR, Hilal N, Lovitt RW, Wright CJ (1999) In: Sorensen TS (ed) Surface chemistry and electrochemistry of membranes. Surfactant science series, Vol 79. Dekker, New York, p 1
3. Binnig G, Rohrer H (1985) Scientific American 253:50
4. Jones RV (1970) Proc IEEE 17:1185
5. Tabor D, Winterton RHS (1979) Proc R Soc London Ser A 312:435
6. Israelaehvili JN, Tabor D (1973) Van der Waals forces: theory and experiment. In: Danielli JF, Rosenberg MD, Cadenhead DA (eds) Progress in surface and membrane science, Vol 7. Academic, New York, p 1
7. Deslattes RD (1968) Appl Phys Lett 15:386
8. Meyer E (1992) Prog Surf Sci 41:3
9. MultiMode™ Scanning Probe Microscope Instruction Manual (1996–97) Digital Instruments Inc., Santa Barbara
10. Butt HJ, Wolff EK, Gould SAC, Northern B, Peterson CM, Hansma PK (1990) J Struct Biol 105:54
11. Amer NM, Meyer G (1988) Bull Am Phys Soc 33:319
12. Meyer G, Amer NM (1988) Appl Phys Lett 53:2400
13. NanoScope III Control System User's Manual, ver 3.0 (1993) Digital Instruments Inc., Santa Barbara
14. Martin Y, Williams CC, Wickramasinghe HK (1987) J Appl Phys 61:4723
15. Magonov SN, Reneker DH (1997) Annu Rev Mater Sci 27:175
16. McLean RS, Sauer BB (1997) Macromolecules 30:8314
17. Zhong Q, Innis D, Kjoller K, Elings VB (1993) Surf Sci 290:L688
18. Hansma PK, Cleveland JP, Radmacher M, Walters DA, Hillner PE, Bezanilla M, Fritz M, Hansma HG, Prater CB, Massie J, Fukunaga L, Gurley J, Elings V (1994) Appl Phys Lett 64:1738
19. Paredes JI, Martínez-Alonso A, Tascón JMD (2003) Microporous Mesoporous Mater 65:93
20. Sheiko SS, Möller M, Reuvekamp EM, Zandbergen HW (1993) Phys Rev B Condens Matter Mater Phys 48:5675
21. Vesenka J, Manne S, Giberson R, March T, Henderson E (1993) Biophys J 65:992
22. MultiModeTM SPM Instruction Manual, ver 4.31ce (1996–1999) Digital Instruments Veeco Metrology Group, Santa Barbara
23. Bowen WR, Hilal N, Lovitt RW, Wright CJ (1998) J Membr Sci 139:269
24. Bowen WR, Hilal N, Lovitt RW, Sharif AO, Williams PM (1997) J Membr Sci 126:77
25. Ducker WA, Senden TJ, Pashley RM (1992) Langmuir 8:1831
26. Atkins DT, Pashley RM (1993) Langmuir 9:2232
27. Li YQ, Tao NJ, Pan J, Garcia AA, Lindsay SM (1993) Langmuir 9:637
28. Rutland MW, Senden TJ (1993) Langmuir 9:412
29. Noy A, Frisbie CD, Rozsnyai LF, Wrighton MS, Lieber CM (1995) J Am Chem Soc 117:7943
30. Overney R, Meyer E, Frommer J, Brodbeck D, Luthi R, Howald L, Guentherodt HJ, Fujihara M, Takano H, Gotoh Y (1992) Nature 359:133
31. Albrecht TR, Dovek MM, Lang CA, Grutter P, Quate CF, Kuan SNJ, Frank CW, Pease RFW (1988) J Appl Phys 64:1178
32. Bowen WR, Hilal N, Lovitt RW, Williams PM (1996) J Colloid Interface Sci 180:350

4 Nodular Structure of Polymers in the Membrane

4.1 Introduction

Important membrane surface properties include the size of nodules and nodule aggregates, the shape of pores, the pore size and pore size distribution, and the surface roughness. In this chapter, the focus will be on nodules and nodular aggregates since AFM seems most suitable for those. Moreover, there is evidence that nodular structure has some relationship to membrane performance.

The phase contrast imaging technique in AFM can distinguished between the amorphous and crystalline phase. A solid formed by the solidification of a chemical and having a highly atomic structure is called a crystal, which has a regular structure and size. On the other hand, a nodule is a mass of polymer molecule agglomerates that are entangled with each other. At a lamellar crystal level, the morphology and crystalline structure are deduced by TEM or X-ray studies. However, the resolution of the AFM can go beyond that easily available with TEM imaging of polymers. At a higher resolution, AFM can give better results and in some cases has revealed unpredicted surface structures. AFM presents surface structures in real space, whereas structural information can be deduced from diffraction data (small angle X-ray scattering or small angle neutron scattering) only in interplay with structural models.

A synthetic polymer may be described as crystalline if it contains regions of three dimensional ordering on atomic (rather than macromolecular) length scales, usually arising from intramolecular folding and/or stacking of adjacent chains. The stacks formed by the folding of chains are called lamellae. Sometimes part of the chain is included in this crystal and part of it isn't. Lamellae are not neat and tidy, but sloppy, with chains hanging out everywhere. The synthetic polymer may consist of both a crystalline and an amorphous region. The crystalline portion is in the lamellae, and the amorphous portion is outside the lamellae. The degree of crystallinity is expressed in terms of a weight fraction or volume fraction of crystalline material.

To examine lamellae and other nanometer-scale structures in polymer materials, it is necessary to achieve high-resolution imaging on the submicron scale. This is easily achieved using Nanoscope® MultiMode™ and Dimension™ 3100 AFMs (Digital Instruments, Veeco Metrology Group, Santa Barbara, CA) under ambient conditions. The necessary prerequisite for high-resolution imaging is a sharp tip. Tapping mode is particularly important for this purpose due to its ability to image soft materials such as most polymers without sample alteration. Low-force imaging or light tapping allows imaging of top surface features with lateral resolution determined by the small tip contact area (2–3 nm). Imaging with elevated forces or hard tapping allows visualization of subsurface structures and differentiation of crystalline and amorphous regions. Height images yield the true three-dimensional topography of the sample surface; the deflection mode is useful for a sharp contrast of the features imaged. Images of the surface of a nodule can expose the lamellar or crystalline phases. The phase contrast imaging technique can be distinguished between the crystalline and amorphous phase.

Nodules are defined as spherical cells with a diameter of a few hundred angstroms that are compacted irregularly at the membrane surface. They can also be observed underneath the membrane surface when a cross-sectional picture is taken. Each nodule contains several tens of thousands of macromolecules. Schultz and Asunmaa were the first to report the observation of nodules on the surface of an ultrathin cellulose acetate membrane by electron microscope [1]. Figure 4.1 shows the picture taken by them. The nodular structure of the membrane surface is clearly seen with an average nodular diameter of 188 ± 3 Å. The same authors also took a picture of an asymmetric cellulose acetate membrane and found that it, too, had a nodular structure. Panar et al. [2] then observed the close monolayer packing of micelles with diameters from 400 to 800 Å when a cross-sectional picture of an asymmetric aromatic polyamide-hydrazide membrane was taken (Fig. 4.2). The top monolayer covers a support layer where the spherical micelles are irregularly packed with void spaces of 75–100 Å. They attributed the formation of the nodules to the micellar structure that was initially present at the surface of the polyamidehydrazide solution.

Nodular structures were found not only in the ultrathin and asymmetric membranes but also at the surface of thin film composite (TFC) membranes. Cadotte reported that nodules were closely packed at the surface of a fully aromatic polyamide TFC membrane prepared by the in situ polycondensation reaction between *m*-phenylene diamine and trimesoyl chloride [3, 4].

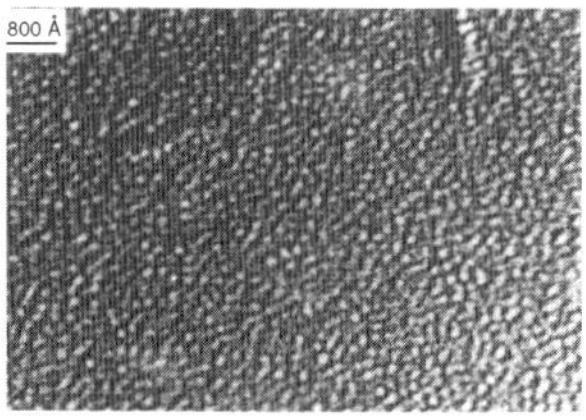

Fig. 4.1. Electron photomicrograph of Pt-C preshadowed carbon replica of the surface of a skin layer of a Loeb-Sourirajan-type cellulose acetate membrane. Reprinted from Polymeric Gas Separation Membranes by R.E. Kesting and A.K. Fritzsche, p 228. Copyright 1993, with kind permission from Wiley

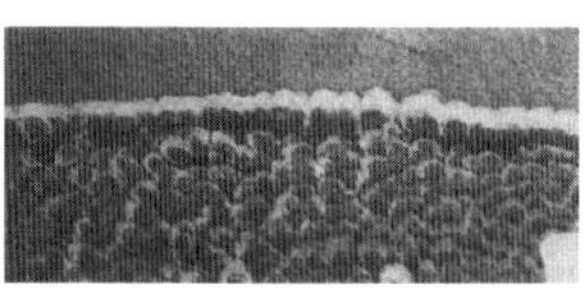

Fig. 4.2. Top edge of cross section of polyamide-hydrazide asymmetric gel membrane taken by SEM. Reprinted from Polymeric Gas Separation Membranes by R.E. Kesting and A.K. Fritzsche, p 248. Copyright 1993, with kind permission from Wiley

Thus, nodular structures are always found at the surface of polymeric membranes. Based on the size of the structural units, Kesting suggested the following four superimposed tiers of structure in asymmetric membranes prepared by the phase inversion technique [5]:

1. Macromolecules
2. Nodules—spherical macromolecular aggregates, approximately 200 Å in diameter, each of which contains several tens of thousands of macromolecules
3. Nodular aggregates—spherical clumps of nodules, from 400 to 1000 Å in diameter
4. Supernodular aggregates—aggregates of nodular aggregates, from 1000 to 20 000 Å (2 μm) in diameter.

Models of different stages are demonstrated in Fig. 4.3. Kesting also discussed the relationship between the nodular and the porous structure of separation membranes.

There are a number of theories for the formation of the nodular structure. Panar et al. [2] attributed nodule formation to the aggregates or micelles that are initially present in the casting solution. Another theory for the generation of nodules is that they are formed as the result of liquid–liquid demixing by nucleation and growth of a polymer-rich phase [6]. However, this theory does not necessarily explain nodule formation in concentrated polymer solutions. There is still another theory that the nodule formation is a surface phenomenon, but this would not explain several layers of nodules in dense film [7]. Broens et al. [8] found nodules in the top layers of poly(phenylene oxide) (PPO) membranes. Nodules were described as structural units in the skin layer formed under fast diffusion processes and originating from gelation or crystallization. According to Ray et al. [9], nodules result from perturbations at the interface of the polymer solution and the coagulation bath. The pertur-

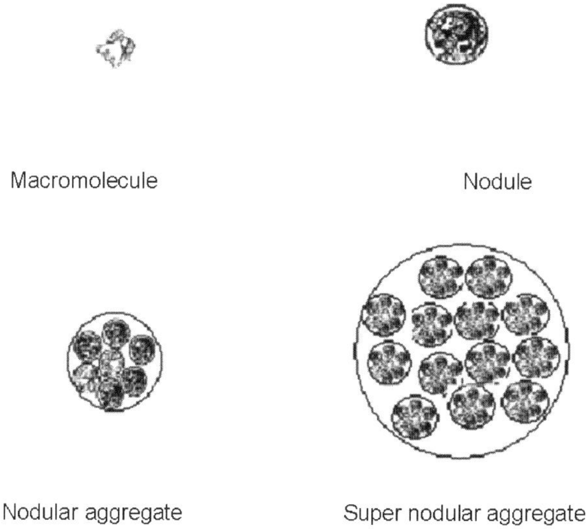

Fig. 4.3. Models as suggested by Kesting

bations are formed due to concentration and temperature fluctuations (Marangoni effect). But if such a surface phenomenon would cause nodule formation, it would not be possible to explain several layers of nodules at the top layer [7].

Kimmerle and Strathmann [10] suggested that the structure was obtained after phase inversion of a polymer solution and dependent on the ratio of the polymer-rich and polymer-lean phases at the moment of phase separation. In the phase diagram, this ratio is determined by the position of the polymer composition at the *tie* line. Reuvers and Smolders [11, 12], by using mass transfer models, showed that immersion of a polymer solution into a bath of a strong nonsolvent caused an increase of the polymer concentration in the top layer. Pinnau [13] prepared polysulfone gas separation membranes from a solution containing a volatile solvent and a nonvolatile nonsolvent. After evaporation of the solvent, the polymer solution was quenched in a nonsolvent bath. In these gas-tight membranes, the top layer consisted of nodules. The author suggested that spinodal demixing was the cause of the nodular structure. In the later stage, collapse of nodules occurred due to capillary forces. Spinodal demixing of a polymer solution resulting in a nodular structure in the top layer of a UF membrane was also reported by Boom et al. [14]. Wienk et al. [7] prepared PES-UF membranes, which had a top layer consisting of nodules, suggesting that the nodular structure was formed due to spinodal demixing. Some researchers have also suggested [15,16] that the nodule formation might depend on the local concentration of polymer. Thus, the formation mechanism of the nodule structure has not yet been elucidated in detail.

There are several reports suggesting that the nodule interiors are denser than interstitial regions [5, 17]. Kawakami et al. [15] prepared 6FDA-APPS dense (evaporation) and asymmetric (dry–wet phase inversion technique [18, 19]) membranes from 2,2′-bis(3,4-dicarboxyphenyl)hexafluoropropane dianhydride (6FDA) and bis[4-(4-aminophenoxy)phenyl]sulfone (APPS) for gas separation. The surface morphology was studied by AFM. They reported that the dry process by evaporation influenced the formation of nodules and that the wet process, by exchange between solvent and nonsolvent at the interface of the coagulation medium, determined the surface roughness of the skin layer.

4.1.1 Nodular Structure on the Membrane Surface: Images of Transmission Electron Microscopy and Scanning Electron Microscopy

Before the invention of AFM, scanning electron microscopy (SEM) and transmission electron microscopy (TEM) were the only tools for surface studies. Both SEM and TEM, however, require sample preparation; i.e., the samples to be subjected to SEM observation should be coated with metals in a vacuum. For the TEM samples, replicas have to be prepared. Such sample preparation could affect the originality of the surface morphology. Moreover, SEM and TEM do not provide clear observations of fine features like nodule boundaries and interstitial regions, which could be obscured by rough topography. For example, it is difficult to estimate the realistic nodule size by SEM due to the thick coating layer of gold [7].

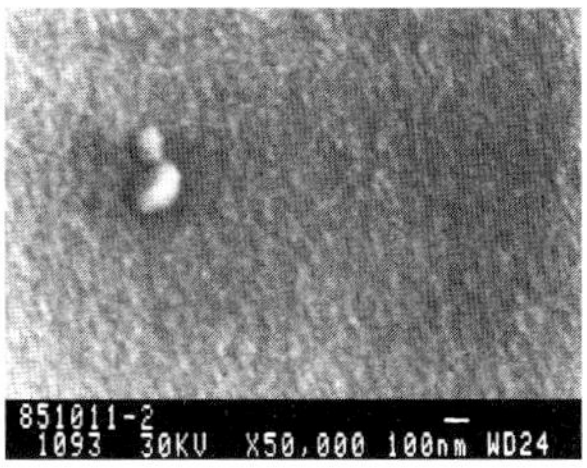

Fig. 4.4. Scanning electron micrograph of the outer surface of a polysulfone hollow membrane spun from formylpiperidine/formamide at ×50 000. Reprinted from Polymeric Gas Separation Membranes by R.E. Kesting and A.K. Fritzsche, p 229. Copyright 1993, with kind permission from Wiley

Figure 4.1 shows the skins of cellulose acetate reverse osmosis membranes [1] (carbon replica of the surface). Yeh and Geil [20] reported similar but smaller structures in poly(ethylene terephthalate). Keith [21] called these structures crystalline nodules. Fritzsche et al. [22] also observed nodules in the surface of asymmetric integrally skinned gas separation membranes of polysulfone (Fig. 4.4) by using the SEM technique. They also revealed the presence of micropores on the dense surfaces.

4.1.2 Studies of Nodules by AFM

Usually, a flat sheet membrane is prepared by spreading a casting solution on a flat surface and evaporating the solvent. A thin polymeric layer that is formed between air and the bulk of the casting solution is called the active or the top layer of the membrane. The performance of the membrane depends largely on the physical or molecular structure of the active layer.

Atomic force microscopy (AFM) was first applied to investigate the polymer surfaces in 1988 shortly after its invention [23]. Today, studies by AFM range from simple visualization of morphology to more advanced examination of polymer structure and properties at the nanometer scale. AFM gives three-dimensional pictures of the surfaces, while other methods, SEM and TEM, do not. AFM is frequently applied to polymer surfaces, principally to reveal morphology, nanostructure, chain packing, conformation, pore size, and pore size distribution at the surface.

As mentioned earlier, nodules are structural units observable at the polymer surface in general and at the membrane surface in particular. The size of a nodule is determined from the cross-sectional profiles of the data along a reference line. An example of the measurement of nodule diameters is shown in Fig. 4.5. The bright sites are nodules and the dark sites are interstitial domains. For each pair of cursors (*pointers*), horizontal and vertical distances are given in the right window. The diameter of the nodules (bright sites), i.e., the maximum width of the cross section of the bright site, can be measured by the help of a pair of cursors, as indicated in Fig. 4.5. By measuring the diameters of a large number of bright sites (at least 25), the average is obtained as the average size of nodules, nodule aggregates, and supernodular aggregates, depending of the size of the bright sites. Similarly, the width of the dark sites, which could be the openings of pores in membranes, can be measured.

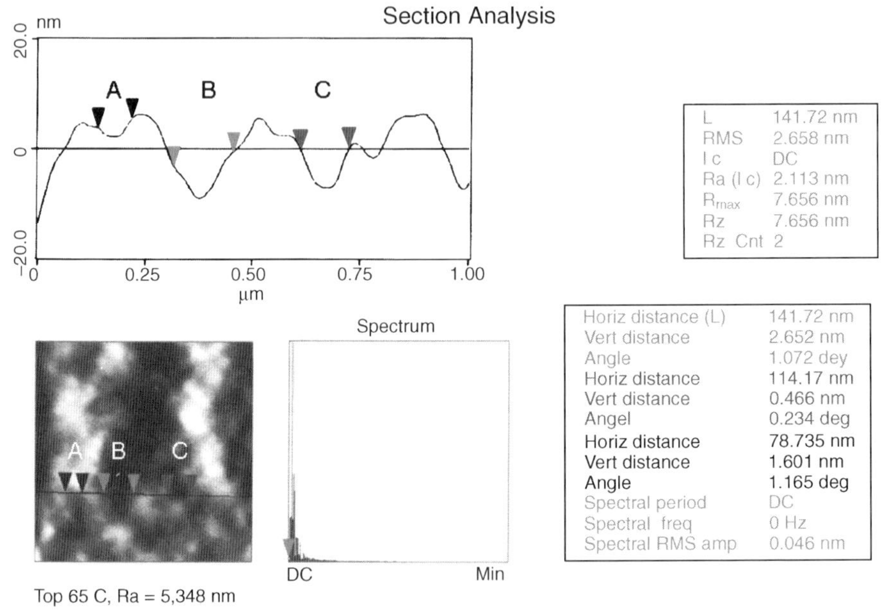

Fig. 4.5. Section analysis of a TM-AFM image: a vertical displacement of the top surface of the dense PPO-TCE membrane. *A*, *B*, and *C* show pairs of cursors for each measurement

4.2 Flat Sheet Membranes

4.2.1 Nodular Structure of the Top Surface

The surface of a poly(phenylene oxide) (PPO) homogeneous membrane was studied by TM-AFM [17] before being used for gas permeation experiments. Membranes were prepared by spreading the polymer solution on a glass plate before the solvent was removed at room temperature by applying a vacuum. Figures 4.6 and 4.7 show three-dimensional AFM images of the bottom and the top surfaces of the homogeneous membrane, respectively. From these AFM images, it seems both surfaces have relatively uniform nodular structures. However, the average diameter of the nodule aggregate at the bottom surface is 137.5 nm, which is twice as large as the average diameter at the top surface (63.8 nm). The mean roughness (see Chap. 3) at the bottom and top surface was 0.75 and 0.37 nm, respectively, when the scan range was 800 nm. The properties of the solvent in the casting solution affect the surface morphology of the membrane. Khulbe et al. [24] prepared PPO dense membranes using different solvents; i.e., carbon disulfide, benzene, tetrachloroethylene, toluene, chlorobenzene and bromobenzene. Figure 4.8 shows the AFM image of the topside of a dense PPO-CS_2 membrane at a scan range of 3 μm. The size of the supernodular aggregates and the roughness parameters of the membrane are summarized in Tables 4.1 and 4.2.

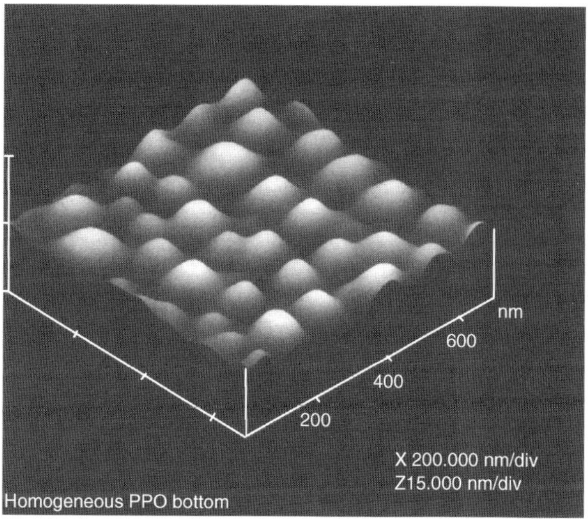

Fig. 4.6. Surface plot of an image by TM-AFM of a homogeneous membrane's bottom surface. Reprinted from [17]. Copyright 1996, with kind permission from Wiley

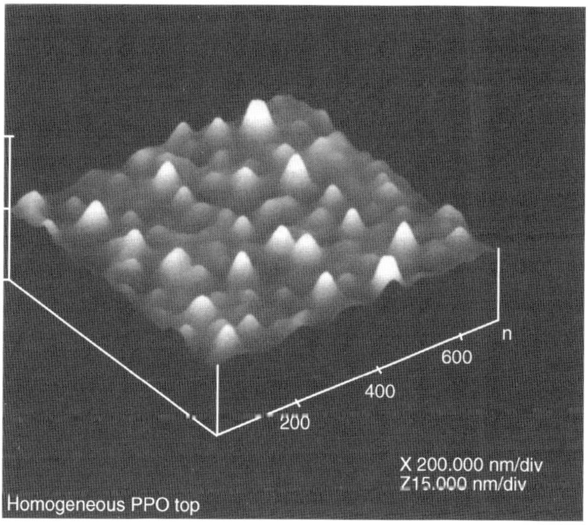

Fig. 4.7. Surface plot of an image by TM-AFM of a homogeneous membrane's top surface. Reprinted from [17]. Copyright 1996, with kind permission from Wiley

As shown in Table 4.1, the mean diameter of the supernodular aggregates is 1.20 μm. Figure 4.9 shows the AFM image of the top surface at a scan size of 1 μm. From Fig. 4.9, it seems that a supernodular aggregate contains smaller units that correspond to nodular aggregates, the mean diameter of which is 43.7 nm. Figure 4.10

Table 4.1. Diameter of the supermodular aggregates at the top surface of the PPO-CS_2 membrane

Scan size (μm)	Mean (μm)	Max. (μm)	Min. (μm)
3	1.2	1.4	0.91

Table 4.2. Summary of the roughness parameters at the top surface of the PPO-CS_2 membrane

Scan size	R_q[a] (nm)	R_a[b] (nm)	R_{max}[c] (nm)
3 μm	12.5	9.9	93.9
1 μm	1.27	1.01	9.45
500 nm	1.24	0.93	9.04

[a] Root mean square (*rms*) of *Z* values
[b] Mean roughness
[c] Mean difference between five highest peaks and five lowest values

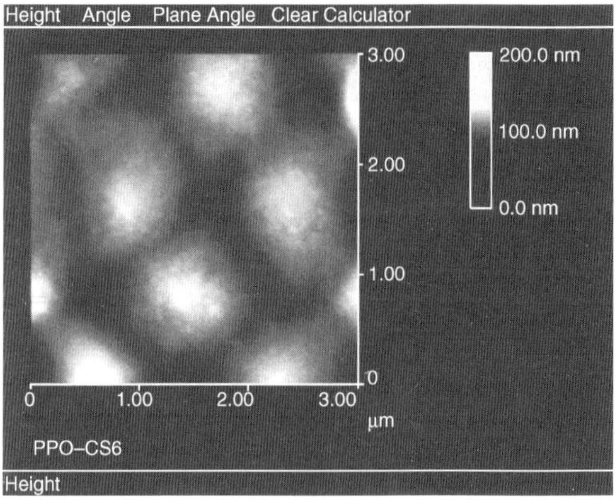

Fig. 4.8. AFM image of the top surface of a dense PPO-CS_2 membrane at 3 μm scan range. Reprinted from [24]. Copyright 1997, with kind permission from Elsevier

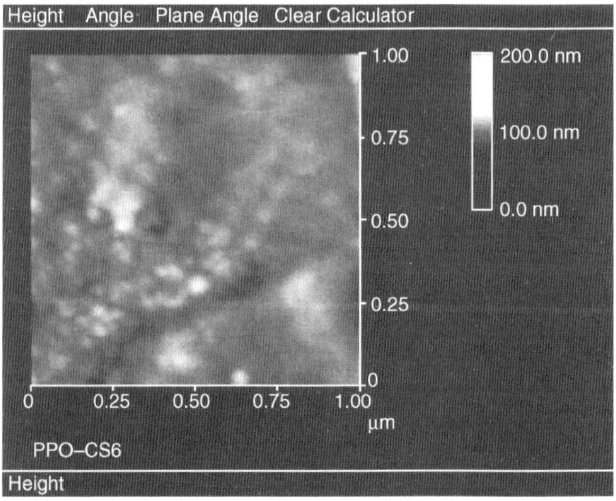

Fig. 4.9. AFM image of the top surface of a dense PPO-CS_2 membrane at 1 μm scan range. Reprinted from [24]. Copyright 1997, with kind permission from Elsevier

shows the three-dimensional AFM image of the top surface of a PPO-CS_2 membrane at a scan size of 500 nm. This picture reveals the detailed structure of one of the supernodular aggregates.

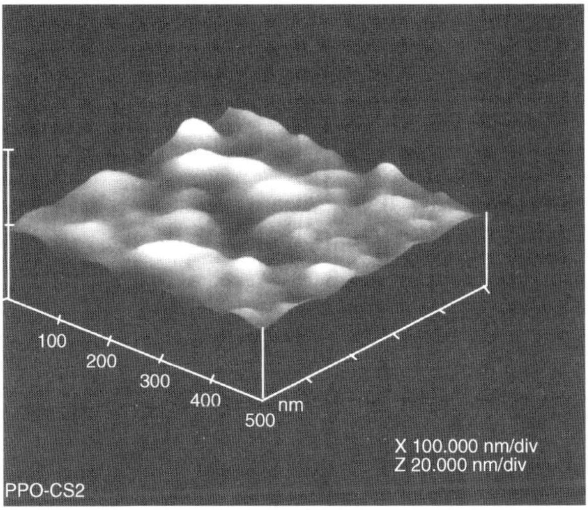

Fig. 4.10. AFM three-dimensional image of the top surface of a supernodular aggregate of a PPO-CS_2 membrane. Reprinted from [24]. Copyright 1997, with kind permission from Elsevier

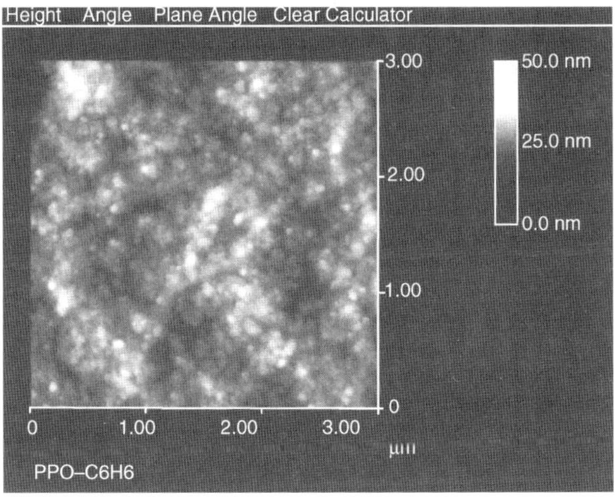

Fig. 4.11. AFM image of the top surface of a dense PPO-C_6H_6 membrane at 3 μm scan range. Reprinted from [24]. Copyright 1997, with kind permission from Elsevier

Figure 4.11 shows the AFM image of a dense PPO-C_6H_6 membrane [24]. The image of the top surface is entirely different from the PPO-CS_2 membrane. Individual nodule aggregates are found to be separate from each other, unlike the CS_2 membrane, where the nodule aggregates were merged to form a large supernodular aggregate. All the membranes prepared with solvents other than CS_2 showed top surface images similar to that of PPO-C_6H_6. Ariza et al. [25] also observed AFM images similar to the top surface of the PPO-C_6H_6 membrane for the commercial membranes supplied by PRIDESA, a Spanish company. Ochoa et al. reported similar pictures for a poly(ether sulfone) (PES) UF membrane [26]. The summary of the nodular aggregate diameters observed at the top surface of the membranes prepared by using different solvents is given in Table 4.3 for the scan size of 500 nm. As well, the mean roughness (R_a) determined at the scan size of 1 μm is summarized in

Table 4.3. No definite conclusions were obtained between the properties of the solvents and the data shown in Table 4.3.

It is interesting to note that the PPO-CS_2 membrane with the supernodular aggregates exhibited the highest selectivity for the CO_2/CH_4 gas pair. Filling the interstitial gap between nodular aggregates may, therefore, lead to higher selectivity of the gas separation membrane.

Asymmetric polyimide membranes with an ultrathin defect-free skin layer were fabricated by the dry–wet process [15]. Composition of casting solution used for the preparation of asymmetric membranes was 12 wt.% polyimide, 55 wt.% methylene chloride, 23 wt.% 1,1,2-trichloroethene, and 10 wt.% butanol. In the dry process (solvent evaporation) the evaporation period was changed from 15 to 600 s, while in the wet process (coagulation process) the coagulation media was methanol. It was possible to control the thickness of the skin layer by controlling the evaporation period. From this AFM study, it was observed that the nodule formation was controlled by evaporation time, while the coagulation media controlled the roughness parameter.

Kwak et al. fabricated polyester high flux RO membranes and studied their surfaces by AFM. A homologous series of thin film composite membranes was prepared by interfacial polymerization of various bisphenols possessing structural variations and trimesoyl chloride (TMC) [27]. The substitution of bisphenol rings with either a methyl or halogen group strongly influenced not only RO rejection and flux but also the features of the resulting aromatic polyester thin-film composite membranes. The AFM observation revealed that, as the number of methyl groups on bisphenol increased, the nodules agglomerated in a more irregular, ambiguous, and smooth approach. On the other hand, halogen-substituted bisphenols resulted in membranes having a surface morphology of larger, fairly uniform, and distinct nodular agglomeration. Moreover, halogen substitution produced membranes with higher salt rejection.

Zhang et al. prepared TFC membranes for nanofiltration by interfacial polymerization of piperazine and trimesoyl chloride on top of polysulfone UF membranes with molecular weight cutoff (MWCO) values of 60 000 Da and water permeabilities of 100–150 $L\,m^{-2}\,h^{-1}\,bar^{-1}$ [28]. Figure 4.12 shows the SEM picture of the sur-

Table 4.3. Summary of the mean nodule diameter and the mean roughness (R_a) of PPO membranes prepared with different solvents

Membrane	Average nodule diameter, scan size 500 nm (nm)	Mean roughness, R_a, scan size 1 μm (nm)
PPO-CS_2[a]	43.7	1.00
PPO-C_6H_6	56.6	1.84
PPO-TCE	49.06	0.17
PPO-$CH_3C_6H_5$	38.95	2.81
PPO-ClC_6H_5	36.7	2.83
PPO-BrC_6H_5	60.76	3.46

[a] Nodule inside the supernodular aggregate

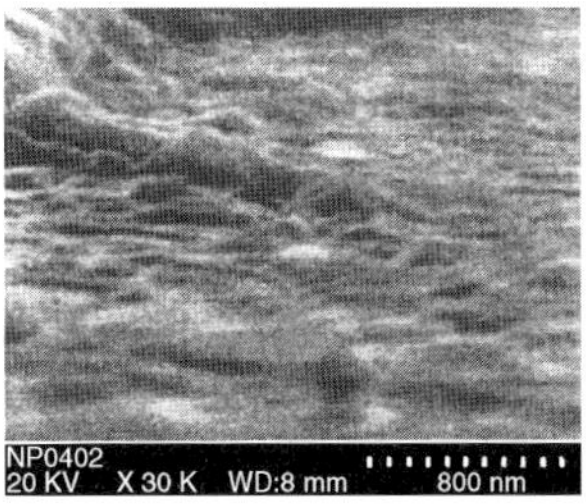

Fig. 4.12. SEM image of typical surface morphology of the NP-1 membrane. Reprinted from [28]. Copyright 2002, with kind permission from Elsevier

face where one can find some protuberance on the membrane. Figure 4.13 shows the AFM images of the membrane's surface in different scales (10 and 15 μm). Both the SEM and AFM images give similar information—that the surface of that particular membrane is uneven and rough with many nodules.

Oh et al. took AFM pictures of TFC nanofiltration membranes, each consisting of a polyamide skin layer formed by in situ polymerization on a polyacrylonitrile (PAN) support layer [29]. PAN support membranes were prepared from casting solutions with different PAN concentrations (10, 15, and 20 wt.%) by the phase inversion technique, followed by treatment with a sodium hydroxide solution. The pictures taken for the top surfaces of these membranes are shown in Fig. 4.14. The figure clearly shows that the surface morphology depends on the PAN concentration.

Hamza et al. preparaed TFC nanofiltration membranes by coating a thin layer of sulfonated poly(phenylene oxide) in hydrogen form (abbreviated as SPPOH, ion exchange capacity 1.93 meq g^{-1} polymer) on top of a porous poly(ether sulfone) (PES) substrate membrane [30]. The coating solution was prepared by dissolving 1 wt.% of SPPOH polymer in chloroform/methanol mixtures of different compositions. It was noted that the intrinsic viscosity of the polymer decreased with an increase in chloroform content in the solvent mixture, indicating that the macromolecules coiled more compactly in a solvent mixture of higher chloroform content. The surface investigation of the membranes revealed their nodule-like structures. Figure 4.15 shows an AFM image of the surface of the SPPOH-PES composite membrane, prepared by using 66 wt.% of chloroform in the solvent. Similar images were obtained for other studied membranes. The ranges of nodule aggregate sizes are given in Table 4.4 for different solvent compositions.

Table 4.4. Ranges of the sizes of nodular aggregates for different compositions of chloroform/methanol solvent mixtures

Composition of solvent (wt.%)		Range of nodule
Methanol	Chloroform	aggregate size (nm)
100	0.0	85–125
82	18	54– 70
58	42	37– 51
34	66	20– 32

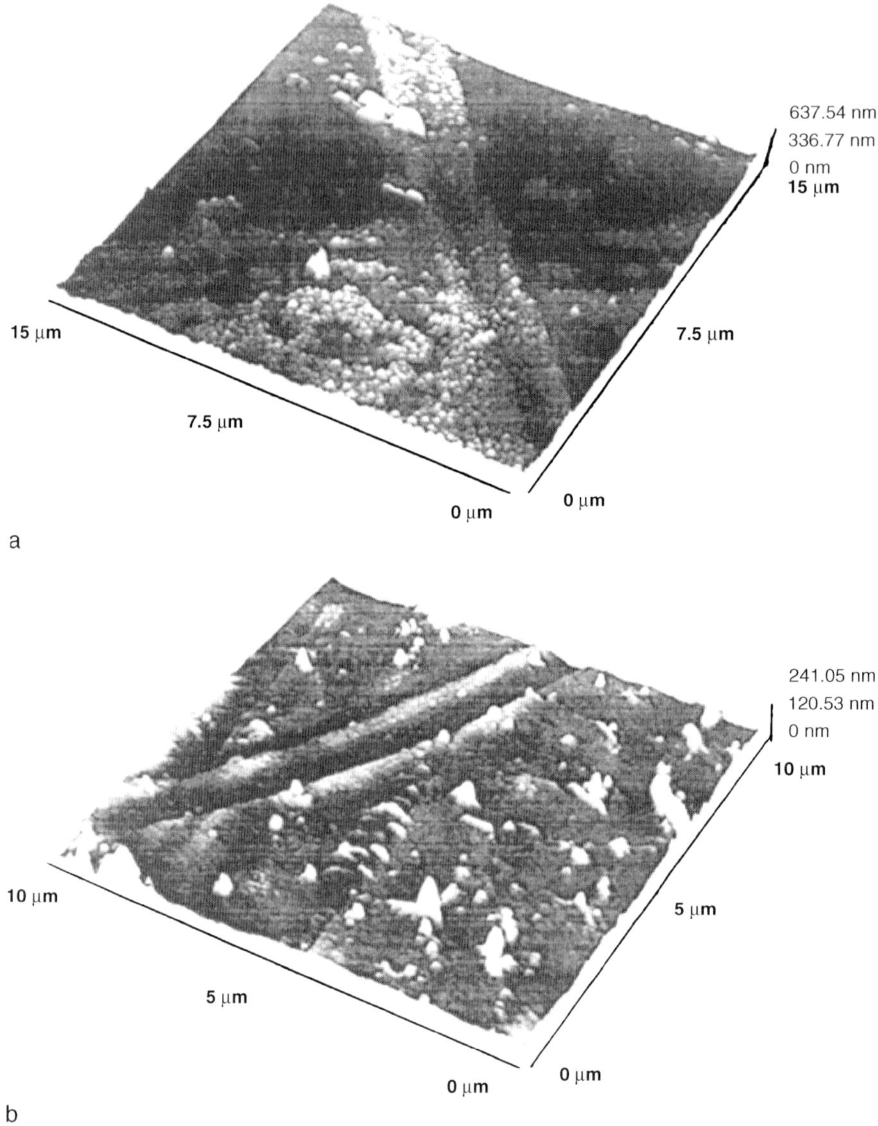

Fig. 4.13. AFM images of typical surface morphologies of the NP-1 membrane. Reprinted from [28]. Copyright 2002, with kind permission from Elsevier

Fig. 4.14a–c. AFM photographs of the PA composite membranes prepared by using modified PAN supports that were prepared from different PAN concentrations: **a** 10, **b** 15, and **c** 20 wt.%. Reprinted from [29]. Copyright 2001, with kind permission from Wiley

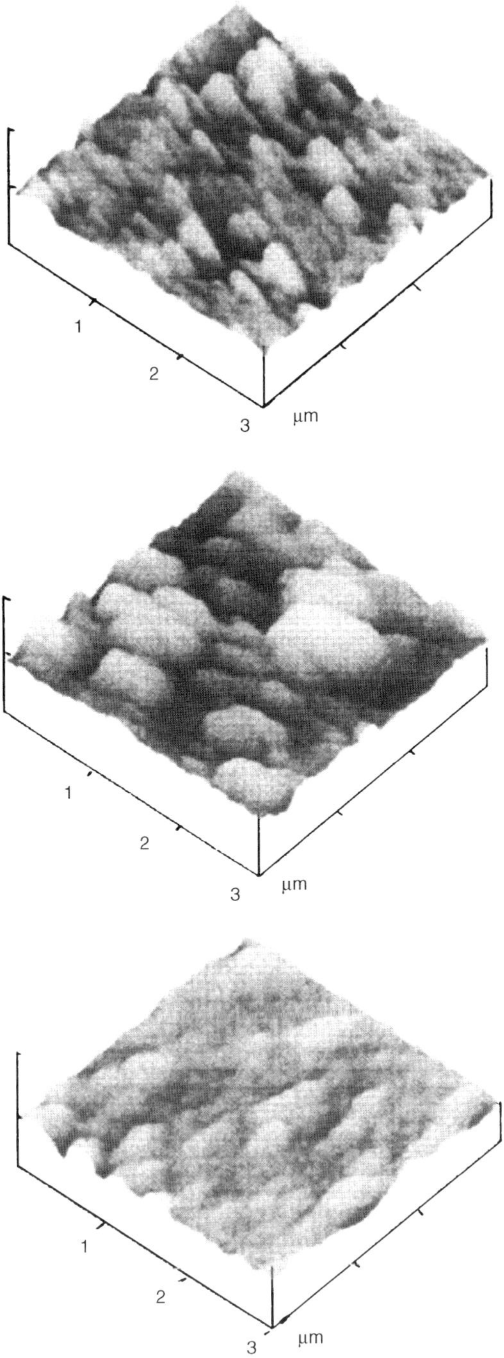

a

b

c

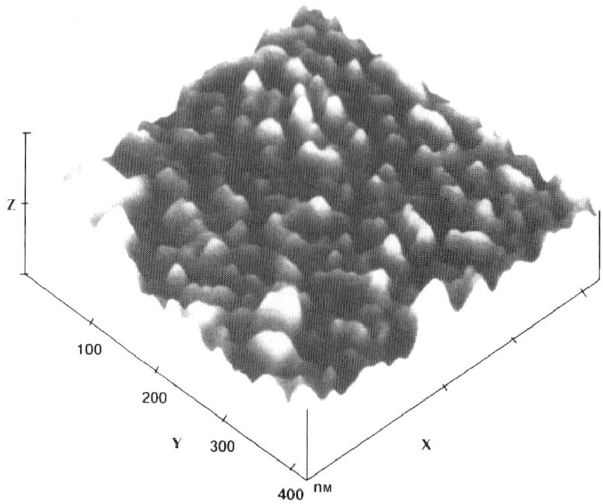

Fig. 4.15. AFM micrograph of the skin layer of SPPOH-PES composite membrane made using 66 wt.% chloroform +34 wt.% methanol as the solvent system for the coating solution. Reprinted from [30]. Copyright 1997, with kind permission from Elsevier

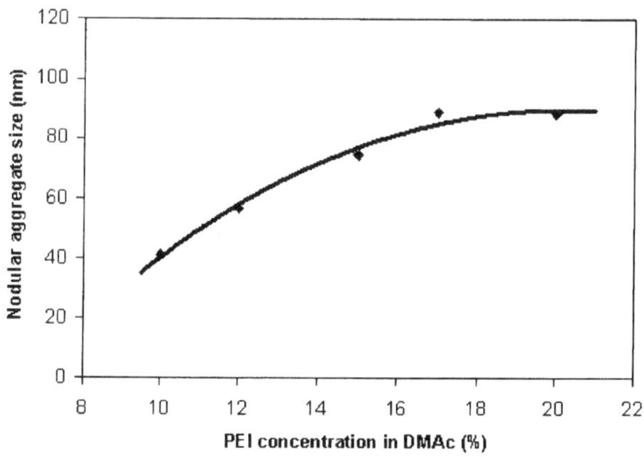

Fig. 4.16. Average nodular aggregate size versus PEI concentration

The size of the nodular aggregate decreased with an increase in chloroform content in the solvent mixture. This means that the compactness of the macromolecular coil in the casting solution was retained after evaporation of the solvent from the solution.

Khayet et al. [31] studied the surfaces of asymmetric poly(etherimide) (PEI) ultrafiltration membranes by AFM. Membranes were prepared by casting mixtures of PEI, hydroxybutyric acid γ-lactone (GBL, γ-butyrolactone) as nonsolvent and N,N-dimethylacetamide (DMAc) at 50 °C. The average size of the nodule aggregates versus PEI concentration is given in Fig. 4.16. From Fig. 4.16, it is clear that the nodule aggregate diameter increases as the PEI concentration increases.

Khayet et al. [32] prepared asymmetric flat sheet membranes for membrane distillation from solutions of poly(vinylidene fluoride) (PVDF) in dimethylacetamide

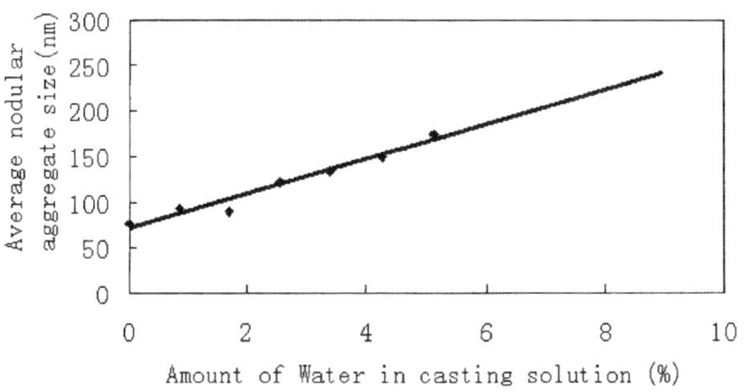

Fig. 4.17. Average nodular aggregate size versus water concentration in casting solution

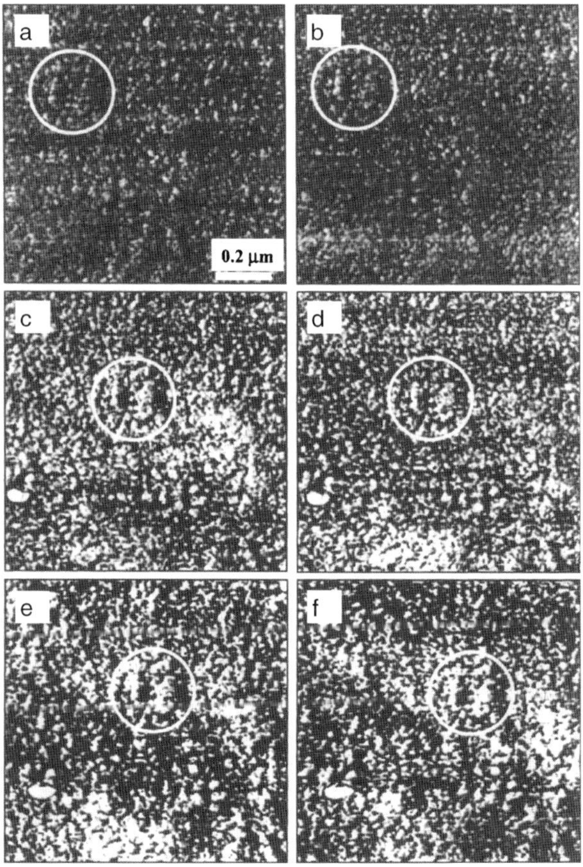

Fig. 4.18a–f. Six 1 μm tapping-mode-phase images of the same area of Nafion at relative humidities of **a** 9, **b** 13, **c** 19, **d** 23, **e** 28, and **f** 33 ± 2% (*z*-scale 30 nm). Reprinted from [33]. Copyright 2000, with kind permission from Kluwer

(DMAc) by the phase inversion technique. The amount of water added to the casting solution as a nonsolvent was changed. It was found that the average nodule aggregate size increased with an increase in the amount of water. At the same time, the mean roughness also increased. Figure 4.17 shows the average nodular aggregate size versus water concentration in the casting solution.

James et al. [33] successfully used tapping mode phase imaging to identify the hydrophobic and hydrophilic regions of Nafion perfluorosulfonate cation exchange membranes. The images support the maximum entropy (MaxEnt) interpretation of a cluster model of ionic aggregation, with spacing between individual clusters ranging from 3 to 5 nm, aggregating to form cluster agglomerates with sizes from 5 to 30 nm. A cluster is defined as an isolated group of white pixels (nodules/nodule aggregate, see Fig. 4.18). The phase images also showed that the number of nodules/nodule aggregate (cluster size) increased with increasing humidity.

Kasper et al. studied dialysis membranes made of Cuprophan® (flat sheet) with AFM and observed differences between modified and unmodified as well as between dry and wet membranes [34]. Modified membranes contained 5, 10, 15, 20, 40, and 100% diethylaminoethylcellulose (DEAE). On the modified Cuprophan® in air as well as under water. These strings may be interpreted as cellulose fibrils, ordered more or less parallel to the membrane production process.

4.2.2 Nodular Structure under the Top Surface: Plasma Treatment

4.2.2.1 Functionalization of Surface by Plasma Treatment

As mentioned in Chap. 2, when a vacuum is maintained inside a tubular reactor and a high frequency field is applied outside, a glow discharge is generated inside the reactor. Plasma that consists of various ions, radicals, electrons, and molecules is formed in the glow discharge. Those species originate from the inserted or residual gas in the reactor and repeat decomposition and recombination while emitting a glow. When a membrane is placed into the plasma, the surface is subject to various changes corresponding to the properties of the plasma.

AFM study of membranes is often combined with the plasma surface treatment. The purpose of plasma treatment in AFM studies is twofold: first, for the modification of the membrane surface and second, for the removal of thin layers, one after another, from the membrane surface to reveal the structure of the membrane beneath. The second type of treatment is called plasma etching.

The capability of plasmas to modify the chemical and physical properties of the surface without affecting the bulk properties of the base material has been advantageous in several cases [35–38]. Either surface modification or thin film deposition can create specific surface chemistries for optimization of membrane performances in separation processes [39]. It is well known that the surface wettability and the adhesion of polymer can be significantly improved by plasma treatment with non-polymer-forming gases. The plasma treatment also leads to the formation of radicals [35] that are promoters of surface cross-linking functionalization.

Vidaurre et al. studied argon plasma-treated (at room temperature) asymmetric polysulfone (PSf) membranes (MWCO 10 000 Da, Danish Separation Systems AS) by AFM imaging [40]. Figure 4.19a and b shows the surface of the polysulfone membranes before and after 10 min of argon plasma treatment, respectively. Comparing the two figures, a remarkable reduction in pore sizes by the plasma treatment can be noticed. Moreover, the roughness parameter, R_a, decreased from 1.6 nm before the

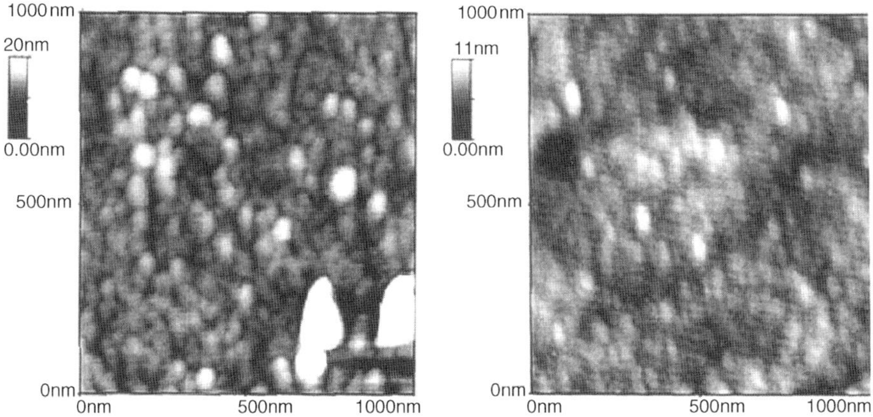

Fig. 4.19. **a** AFM image of the top surface of the PSf asymmetric UF membrane (substrate), R_a 1.6 nm. **b** AFM image of the top surface of the PSf modified by argon plasma at 5 W, 10 min, R_a 1.2 nm. Reprinted from [40] with kind permission from Materials Research, Universidade Federal de Sao Carlos

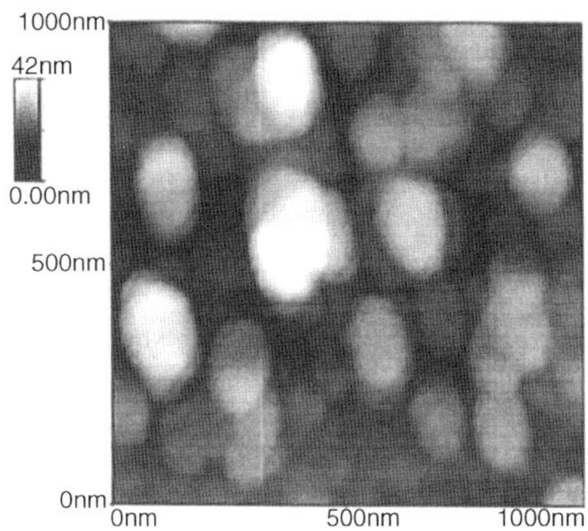

Fig. 4.20. AFM top view image of the surface of a PSf membrane modified by argon plasma at 15 W, 20 min, R_a 5.2 nm. Reprinted from [40] with kind permission from Materials Research, Universidade Federal de Sao Carlos

treatment to 1.2 nm after the treatment. The reduction in the roughness parameter can be interpreted as a consequence of the reduction of the pore sizes due to cross-linking. Interestingly, the gas permeability kept decreasing until the plasma treatment period reached 20 min.

In this context, it may be worth noting Shimomura et al.'s work [41]. They have demonstrated that argon-treated PAN membranes had a thin layer insoluble in dimethylformamide (DMF) solvent. This suggests that a certain number of cross-links were introduced by plasma treatment. Going back to Vidaurre et al.'s work [40], Fig. 4.20 shows the AFM picture of a PSf membrane plasma treated for 20 min. Large nodules (bright spots) surrounded by large interstitial spaces (dark areas) are present on the surface. The latter is most likely the pore entrances. Cross-links formed by 10 min treatment were broken and pores were enlarged. Corresponding to the AFM pictures taken at various plasma treatment periods, nitrogen permeability showed a minimum at 10 min of plasma treatment (Fig. 4.21).

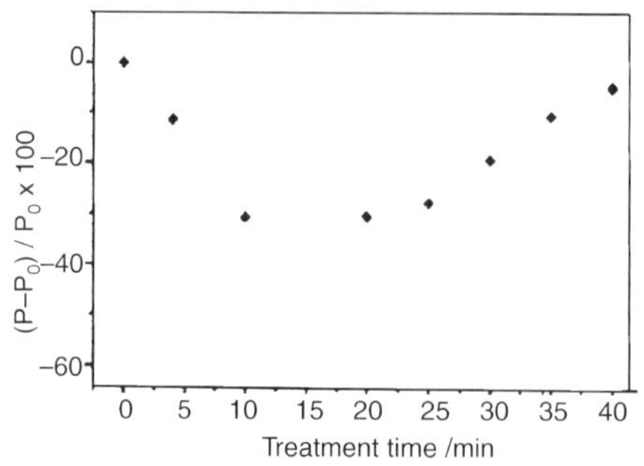

Fig. 4.21. Nitrogen permeability through UF PSf membranes after argon plasma treatment at 5 W. Reprinted from [40] with kind permission from Materials Research, Universidade Federal de Sao Carlos

In another study, Vidaurre et al. [42] modified the surface of PSf membranes by plasma treatment under ammonium gas. The chemical and physical characterization of the membranes before and after the plasma treatment was done by means of AFM, SEM, and XPS. The permeation rate of pure gases (N_2 and CO_2) was measured using a conventional gas permeation cell at room temperature. It was clearly observed that the ammonium plasma treatment affected the surface morphology of the PSf asymmetric membranes. Figure 4.22 shows a three-dimensional AFM image of the top surface, which features a nodular structure with interconnected cavity channels running between the agglomerated nodules. This observation is similar to that of Kim et al. [43]. Figure 4.23 shows, on the other hand, an AFM image of the top surface after the ammonium plasma treatment [42]—cracks formed on the surface. More cracks were formed when the membrane was exposed to higher powers even for shorter times, indicating that this process depends mainly on the total amount of energy deposited to the system during the plasma treatment. Higher gas permeability was observed after plasma treatment, and the increase in permeability was highest at the highest power input. Shimomura et al. also observed the cracks when the PSf polymer surface was exposed to a helium plasma treatment at high power for a period longer than 10 min [41].

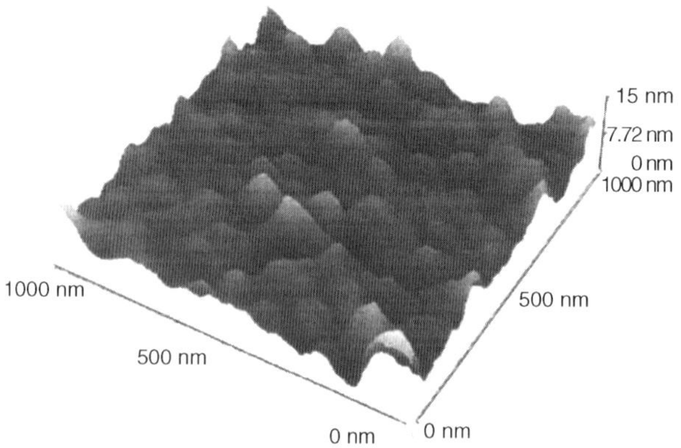

Fig. 4.22. AFM image of the top surface of the PSf asymmetric UF membrane (substrate) (three-dimensional view of a 1 × 1 μm area). Reprinted from [42]. Copyright 2001, with kind permission from Elsevier

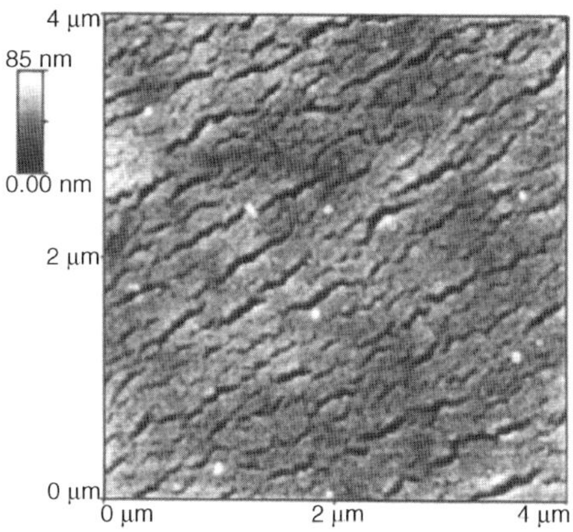

Fig. 4.23. AFM image of the top surface of the PSf membrane modified by ammonium plasma, 15 W, 4 min, 4 × 4 μm top view. Reprinted from [42]. Copyright 2001, with kind permission from Elsevier

Dreux et al. studied the effect of CF_4 and CO_2 plasma treatment on the barrier properties of polyamide 12 (PA12, ATOFINA, Serquigny, France) toward permeant molecules of opposing characters, i.e., water and toluene [44]. While CF_4 treatment made the surface more hydrophobic, CO_2 treatment made it more hydrophilic. The surfaces were studied by AFM and XPS.

The CF_4 plasma treatment led to a decrease in the permeability of both water and toluene. With the CO_2 treatment, on the other hand, water permeability increased and toluene permeability decreased. Figure 4.24 shows the AFM image of the untreated PA12 surface for different scan sizes. Figures 4.25 and 4.26 show the AFM images of CF_4 and CO_2 plasma-treated PA12 membranes, respectively. Obviously, the surface changed after plasma treatments. A denser layer was formed by the plasma treatment.

Also, from the contact angle measurements and X-ray photoelectron spectroscopy analysis, it was concluded that a *layer* was formed at the membrane surface by the plasma treatment. The newly created layers were inhomogeneous in structure and thickness, and it was not possible to estimate the thickness of the layers. In any case, the morphology of the surface of the polymer changed due to the plasma treatment, which was due to various reactions such as functionalization and cross-linking, but at the same time, degradation took place. For the CF_4 plasma-treated surface, the smaller structure of spherules (possibly nodule aggregates) in the layer could be the result of competition between fluorination and degradation reactions [44]. For CO_2 plasma-treated surfaces, nodular structures also appear. At a higher scan, some dark round spots (pinholes) were observed, which could be due to degradation [45].

Compared to CF_4 plasma treatment, CO_2 plasma treatment led to a change in the morphology toward the less homogeneous, and the change covered practically the whole surface. In both cases, the alteration of the topography cannot be the result of deposition because CF_4 and CO_2 are known not to be polymerized by plasma [46, 47]. Rather, the alteration is a modification of the surface by substitution followed by cross-linking and degradation phenomena.

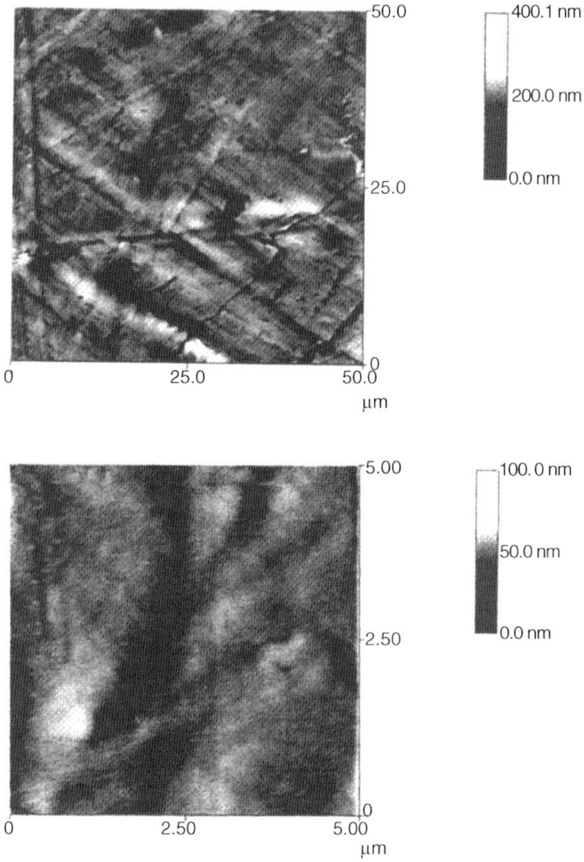

Fig. 4.24. Images obtained by AFM of the untreated PA12 surface at different scales (50 × 50 μm^2 and 5 × 5 μm^2). Reprinted from [44]. Copyright 2003, with kind permission from Springer-Verlag

Finot et al. [48] deposited a plasma-polymerized layer on commercial membranes made of cellulose ester (MF25 Millipore, mean pore size 25 nm, porosity 70%) by HMDSO (linear alkylsiloxane hexamethyldisiloxane) plasma polymerization. Then, the membranes were further post-treated with CF_4/Ar plasma. In the first plasma treatment with HMDSO, polymerization took place and the product polymer was deposited on the membrane surface. In the second treatment, CF_4/Ar plasma produced F· and CF_x· radicals, which etched the surface of the deposited layer, removing (by exodiffusion) reducing agents such as H_2 and CH_x and replacing them. In the first plasma treatment, three samples were prepared and labeled as S (soft), M (medium) and H (hard), corresponding to the three conditions of low, medium, and high values of $V/F'M$, where V is applied voltage, F' the monomer flow rate, and M the molecular weight of HMDSO (146 g mol^{-1}). The details of the plasma polymerization conditions are given in Table 4.5. No matter what the conditions were, the thickness of the deposited polymer layer was approximately 500 nm.

The second plasma treatment was carried out in a special reactor (capacitively coupled reactor) devoted to this reaction. Figure 4.27 shows the morphological char-

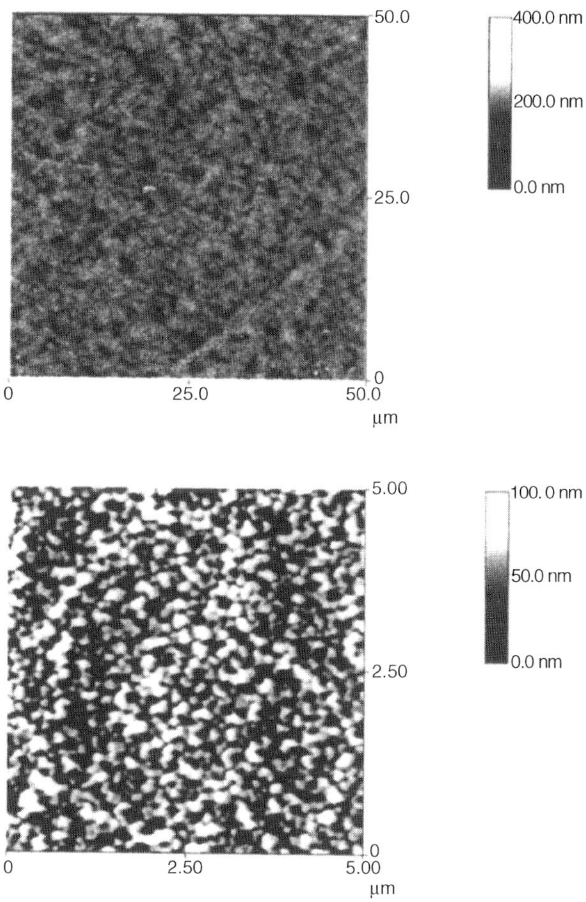

Fig. 4.25. Images obtained by AFM of the CF_4 plasma-treated PA12 surface (50 W, 10 seem, 10 min) at different scales (50 × 50 μm^2 and 5 × 5 μm^2). Reprinted from [44]. Copyright 2003, with kind permission from Springer-Verlag

Table 4.5. Conditions of HMDSO plasma deposition

Samples	V (volt)	F' (mol min^{-1})	$V/F'M$ (V min g^{-1})
S	50	3.8×10^{-4}	1000
M	50	10^{-4}	3000
H	100	7.7×10^{-5}	8000

acteristics of the surfaces. In the figure, a, c, and e correspond to the surfaces of membranes after the first plasma treatment under the S, M, and H conditions, respectively, while b, d, and f correspond to the surfaces after the second treatment of S, M, and H samples (SF, MF, and HF), respectively. Pictures a, c, and e show that the surfaces are different depending on the conditions of the first treatment. The sizes (in diameter) of nodules, nodule aggregates, and supernodular aggregates are summarized in Table 4.6.

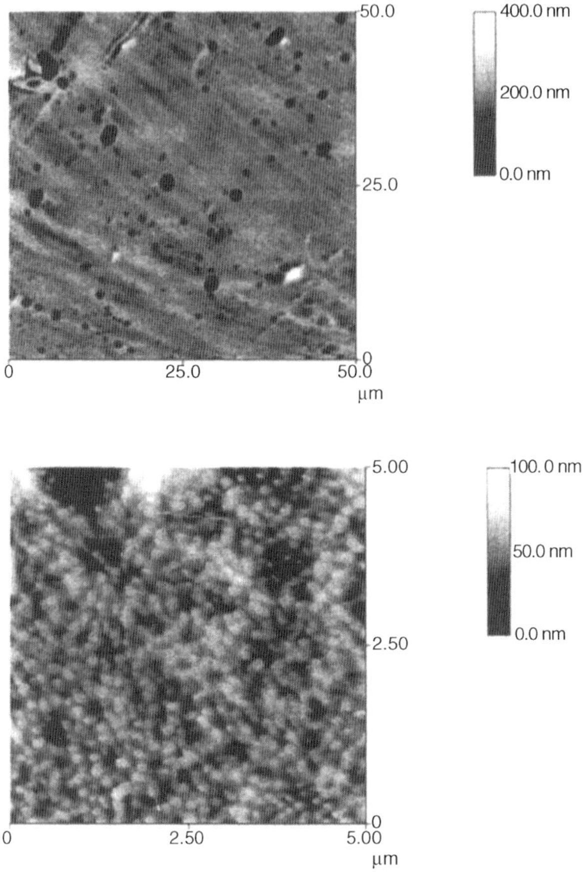

Fig. 4.26. Images obtained by AFM of the CO_2 plasma-treated PA12 surface (50 W, 10 seem, 10 min) at different scales ($50 \times 50\ \mu m^2$ and $5 \times 5\ \mu m^2$). Reprinted from [44]. Copyright 2003, with kind permission from Springer-Verlag

Table 4.6. Apparent average diameters of the particles constituting the various plasma-treated cellulose ester membranes

Samples	Nodules (nm)	Nodule aggregates (nm)	Supernodular aggregates (nm)
S	20 ± 4	420 ± 140	0
SF	13 ± 5	200 ± 60	600 ± 200
M	16 ± 4	340 ± 60	0
MF	11 ± 6	240 ± 100	700 ± 300
H	12 ± 2	140 ± 60	620 ± 140
HF	12 ± 7	150 ± 40	460 ± 100

The diameter of the nodules appears to be 20 and 16 nm for S and M membranes, respectively. For the H membrane, supernodular aggregates of 620 nm contained 6–8 nodule aggregates of 140 nm (Fig. 4.27). In addition to their small diameter (12 nm), the SiO_2 nodules are more discernable than those of M and S membranes.

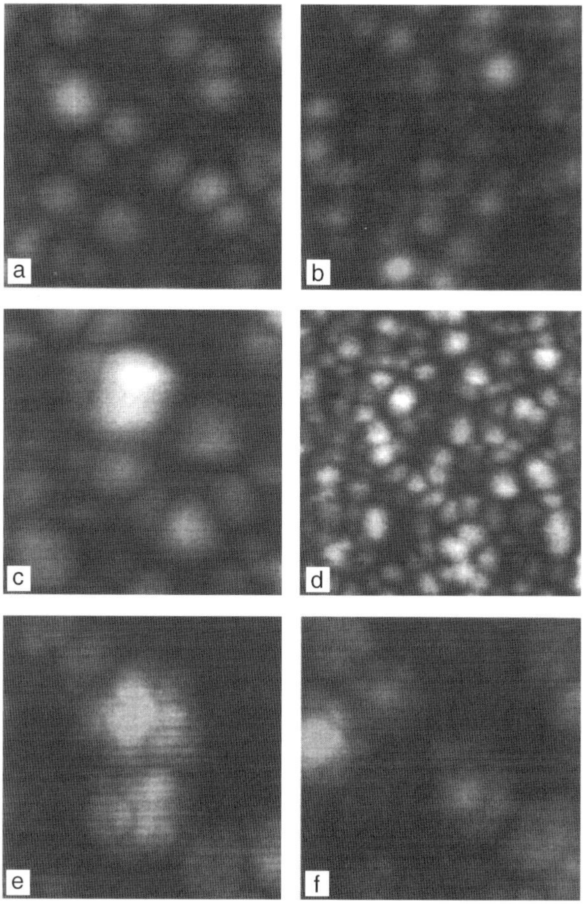

Fig. 4.27a–f. Height images of membranes before and after fluorination treatment (scan size 2×2 μm): **a** S, **b** SF, **c** M, **d** MF, **e** H, and **f** HF. Reprinted from [48]. Copyright 2002, with kind permission from Elsevier

4.2.2.2 Plasma Etching

Another technique is called plasma etching, or sometimes plasma ablation. When a membrane surface is exposed to high-energy particles in plasma, macromolecules are destroyed and removed from the surface as small molecules. The longer the surface is exposed to the plasma, the deeper the removed surface layer becomes. Investigation of AFM surface images after different exposure periods reveals the structural change of macromolecules toward the depth direction of the membrane. This technique is very important in AFM imaging in order to know the profile of the nodule sizes in the depth direction, since, unlike SEM, cross-sectional images are seldom obtained by AFM.

Of course, SEM pictures can also be taken after different periods of plasma etching. This technique was used by van't Hoff [49], Fritzsche et al. [22, 50, 51],

Fritzsche [52], and Weigel et al. [53] on different polymeric membranes. Oxygen plasma ablation technique was shown to be useful for studying the structure of hollow fiber membranes [51]. Fritzsche et al. [51] showed by plasma ablation of a polysulfone hollow fiber membrane that pore size and porosity of the internal supporting matrix increased with sequential removal of the outer layers of the hollow fiber membranes. By oxygen plasma ablation technique, they also reported that the active separating layer depends on the ingredients of the solvent (aliphatic acids + *N*-methyl pyrrolidone) used to make the spinning dopes. In another study, Fritzsche et al. [22], by treating asymmetric polysulfone hollow fiber with plasma, reported that the enhanced free volume with a graded density skin exists in the effective separating layer as well as in the membrane interior. Weigel et al. [53] treated acrylonitrile polymeric membranes with a low-temperature plasma and discussed the relationship between the structural changes and the parameters of plasma treatment.

Khulbe and Matsuura characterized the dense PPO membrane etched by oxygen plasma [54] and studied the structural changes in the depth direction by AFM. Carbon disulfide was used as a solvent when the dense membrane was prepared. It should be recalled that unlike other solvents, carbon disulfide produced a dense PPO membrane, and supernodular aggregates could be observed on its surface.

Figure 4.28 reproduces the AFM image of the top surface layer of an unetched PPO-CS_2 membrane corresponding to the scan size of 5 µm. The bright spots seem most likely to be supernodular aggregates with an average diameter of 0.976 µm, which is slightly less than the one reported earlier [24]. This change could be due to some environmental change (temperature, humidity, etc.) during the fabrication of the membrane. Figure 4.29 shows the AFM image of the top layer of the PPO-CS_2 membrane after oxygen plasma etching of 800 s. While large bright spots (supernodular aggregates) can still be identified, it is obvious that they consist of a number of smaller objects, which are considered to be nodules/nodule aggregates. It suggests that a thin polymer layer that was filling the spaces between individual nodules/nodule aggregates was removed by plasma etching. Even though it is not very clear from Figs. 4.28 and 4.29 alone, the dark area between the bright spots also increased as the plasma etching period increased. Figure 4.30 shows the AFM image of the top layer of the PPO-CS_2 membrane after oxygen plasma etching of 3200 s.

Table 4.7 shows that the average width and length of the dark area both between the supernodular aggregates (A) and between the nodular aggregates (B) increase with an increase in etching time. Similar results were also observed with a PPO-TCE membrane on oxygen plasma etching. Based on the above observations, the morphological changes in the depth direction of a dense PPO membrane were schematically illustrated by Fig. 4.31. At the top-most layer are supernodular aggregates that contain nodular aggregates. Supernodular aggregates are covered by a thin layer of polymer so that the individual nodular aggregates are not identifiable. Most likely, the space between the supernodular aggregates is also filled with polymer, since otherwise, the selectivity of the membrane cannot be as high as experimentally observed. Going downward in the depth direction, the distance between the supernodular aggregates increases. Unlike the top surface, the amount of polymer between the supernodular

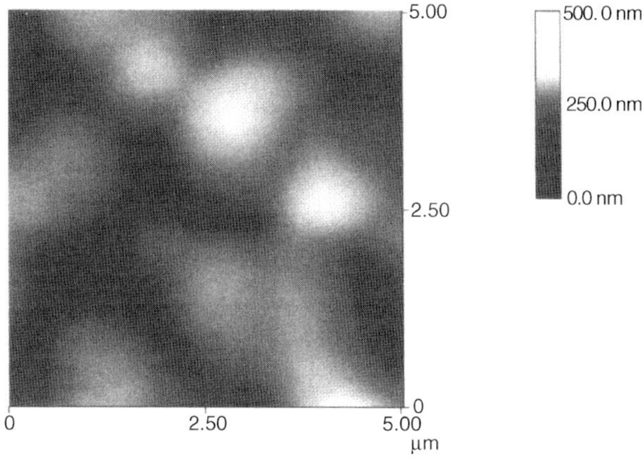

Fig. 4.28. AFM image of the top surface of the PPO-CS_2 membrane at a 5-μm scan size. Reprinted from [54]. Copyright 2000, with kind permission from Elsevier

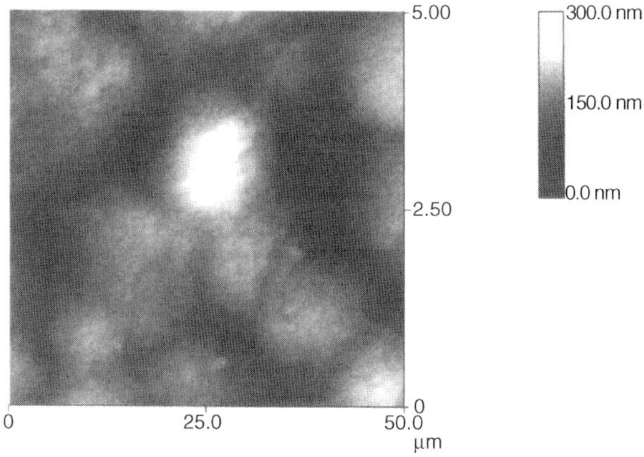

Fig. 4.29. AFM image of the top surface of the PPO-CS_2 membrane after etching for 800 s. Reprinted from [54]. Copyright 2000, with kind permission from Elsevier

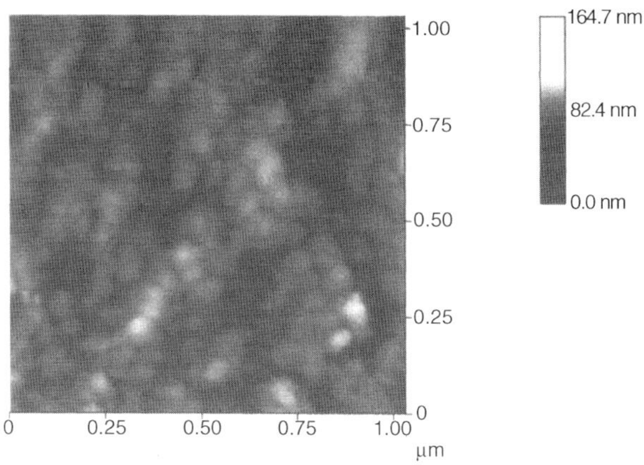

Fig. 4.30. AFM image of the top surface of the PPO-CS_2 membrane after etching for 3200 s. Reprinted from [54]. Copyright 2000, with kind permission from Elsevier

aggregates is less. The distance between the nodular aggregates also increases in the depth direction. The size of the supernodular aggregates increases in the depth direction as well, although the change is not shown in Fig. 4.31. Fujii et al. [55] observed similar morphology at the cross section of their PMMA B-1 (polymethyl methacrylate) hollow fiber by using the field emission scanning electron microscopy (FE-SEM) technique. Figure 4.32 shows a cross section of the PMMA hollow fiber. The inside surface layer is composed of compactly packed polymeric spheres. In the middle layer of the wall, the polymeric spheres are larger and more loosely packed than in the in-

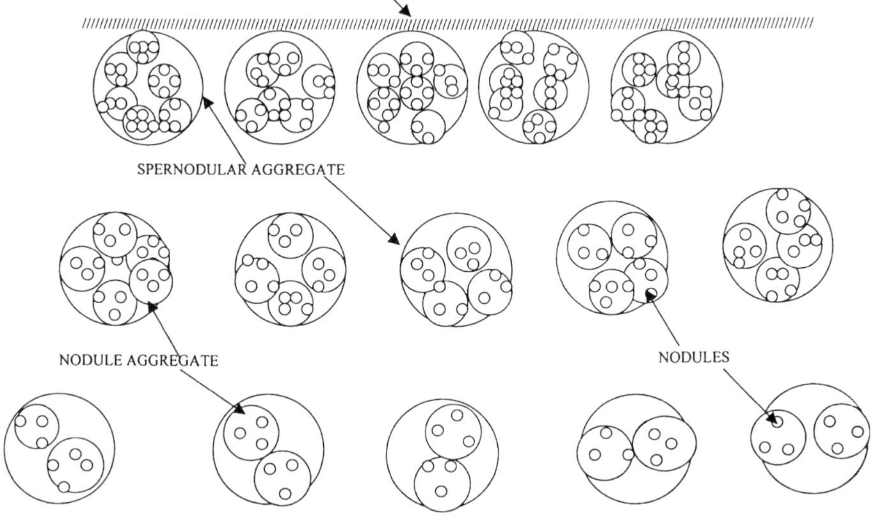

Fig. 4.31. Arrangement and size of supernodular aggregates and nodular aggregates in dense PPO-CS_2 membrane. Reprinted from [54]. Copyright 2000, with kind permission from Elsevier

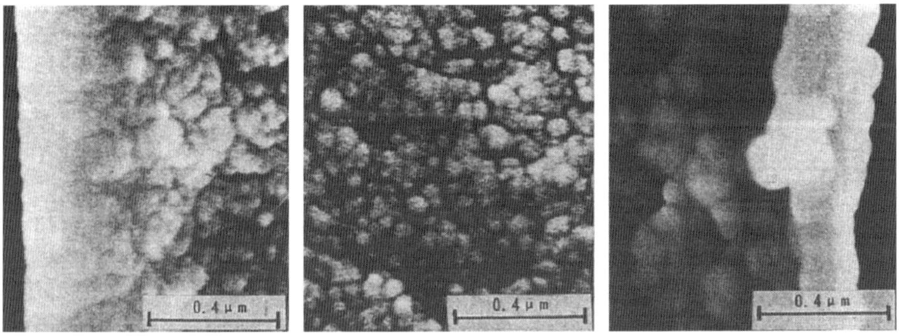

a Insider layer b Middle layer c Outside layer

Fig. 4.32. Cross section of PMMA B-1 hollow fiber membrane (FE-SEM). Reprinted from [55], with kind permission from the Society of Polymer Science, Japan

Table 4.7. Mean width and length of the gap between supernodular aggregates after various etching times, at 5-μm scan range (A), and at 1-μm scan range (B) in the PPO-CS_2 membrane

Plasma etching time	A Width (nm)	Length (μm)	B Width (nm)	Length (nm)	Decrease in thickness of the membrane [a] (μm)
0	532	1.24	49	59	0.62
800	738	1.36	46	81	0.62
1600	882	1.09	43	100	0.91
2400	924	1.23	73	229	1.31
3200	1090	1.44	125	230	1.95
3600	1075	1.16	125	191	2.68
4000	1150	1.07	157	309	3.10

[a] Original thickness of the membrane was 22 μm.

side. The outside surface layer is composed of much larger spheres than the others. Boom et al. studied the cross section of a PES–PVP membrane by SEM and reported that on the surface of the hollow fiber, nodule size was estimated to be ~ 50 nm [14]. In the cross section, the nodular size increased from 50 nm (at 0.5 μm from the surface) to 100 nm (at 5 μm below the surface). Below the top layers, the membrane had an open pore structure.

4.3 Hollow Fiber Membranes

Hollow fibers have been used since the 1960s in many applications such as reverse osmosis, ultrafiltration, membrane gas separation, artificial organs, and other medical purposes. There are several advantages to hollow fibers over the flat sheet membranes; the most important is their high surface-to-volume ratio. The use of hollow fibers has become popular in many industrial sectors since Mahon first patented the hollow fiber membranes [56]. The morphology and performance of hollow fibers are complex functions of many parameters involved in their manufacturing. McKelvey summarized the effect of spinning parameters on the macroscopic dimensions of hollow fibers [57].

Earlier attempts in the study of hollow fiber morphology are based on cross-sectional pictures taken by SEM, by which the asymmetric structure of the fiber membranes was clearly seen. In contrast to the SEM, morphological studies of hollow fibers by AFM are mostly based on the image of the fiber surface, either on the inside or the outside. A cross-sectional picture has seldom been taken.

Chung et al. demonstrated the effect of the shear stress working from the spinneret wall to the outermost surface of the spinning dope [58]. The hollow fibers were spun with no air gap so that the surface morphology could be frozen in the coagulation bath immediately after the fibers extruded from the spinneret. Then, the AFM image of the outer surface of polysulfone (PSf) hollow fibers was obtained.

The AFM image revealed that the nodules in the outer skin appeared random at low shear rates but formed bands that were aligned in the direction of dope extrusion when the shear rate was increased. Figure 4.33 shows the AFM image of the outer layer when the shear rate was 1305 s^{-1}, while Fig. 4.34 corresponds to the shear rate of 11 066 s^{-1}. Chung et al. [58] claimed that these AFM images supported their hypothesis, that *the higher shear stress at the spinneret may result in a hollow fiber UF membrane with a denser skin due to the greater molecular orientation and the closer package of molecules.*

Table 4.8 shows the effect of shear rate on the roughness of the outer surface. It is clear that there is a critical shear rate at 3585 s^{-1}, below which the roughness decreases with an increase in shear rate before it levels off. Since their study was focused on the

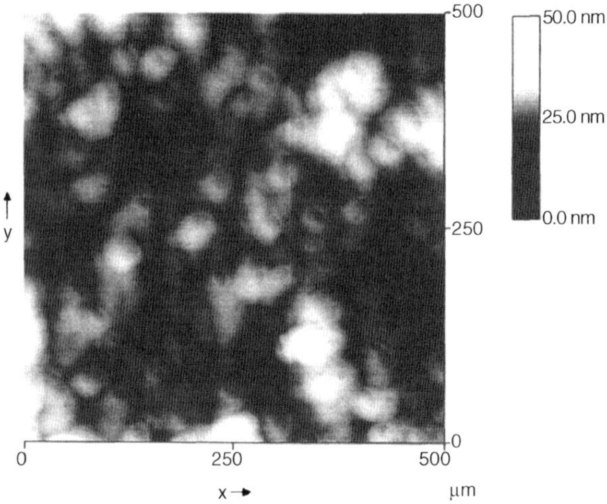

Fig. 4.33. Top view AFM image of outer surface of hollow fiber UF membrane with shear rate 1305 s^{-1}. Reprinted from [58]. Copyright 2002, with kind permission from Elsevier

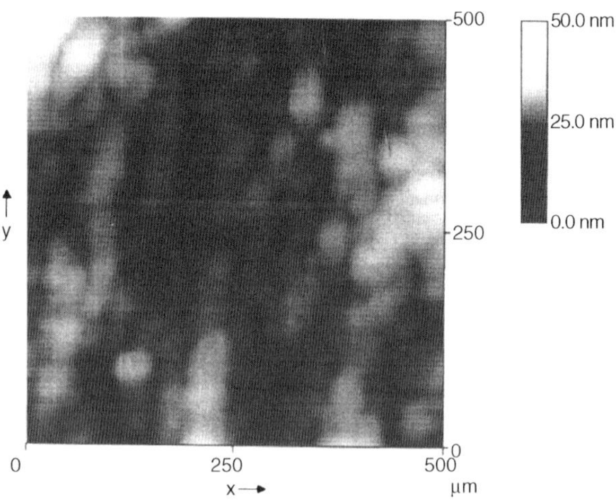

Fig. 4.34. Top view AFM image of outer surface of hollow fiber UF membrane with shear rate 11 066 s^{-1}. Reprinted from [58]. Copyright 2002, with kind permission from Elsevier

Table 4.8. Effect of shear rate on roughness of the outer surface of hollow fiber membranes

Fiber ID	Shear rate (s^{-1})	R_a[a] (nm)	R_q[c] (nm)	R_z[d] (nm)	Dimension of nodules in x-direction (nm)	Dimension of nodules (fiber) in extrusion direction (nm)
1	1305	2.54 (0.19)[b]	3.08 (0.23)	10.4 (1.7)	75.2 (7.2)	74.9 (7.0)
2	2279	1.86 (0.15)	2.33 (0.18)	6.27 (0.45)	64.7 (6.2)	65.0 (6.4)
3	3585	1.52 (0.12)	1.84 (0.15)	5.23 (0.37)	59.6 (5.2)	58.7 (4.8)
4	8235	1.35 (0.11)	1.67 (0.13)	5.04 (0.34)	56.6 (4.8)	56.1 (4.8)
5	11 066	1.21 (0.11)	1.58 (0.12)	4.92 (0.30)	54.7 (4.4)	55.1 (4.6)

[a] R_a, mean roughness
[b] Standard deviation (data inside parenthesis)
[c] R_q, root mean square of Z values
[d] R_z, 10-point mean roughness

effect of the shear stress working on the outer surface of the hollow fiber, they have not observed the inner surface.

Kapantaidakis and Koops [59] and Kapantaidakis et al. [60] studied the formation and gas permeation properties of hollow fiber membranes based on poly(ether sulfone)/polyimide blends of three compositions (80/20, 50/50, and 20/80 weight ratio). They reported that the air gap affected both membrane structure and permeation properties in the dry–wet spinning process, where the bore fluid (NMP/water 80/20) was supplied from the central tube of the spinneret to the center of the fiber at a flow rate of 1.12 mL min^{-1}. The fiber traveled the air gap before entering the coagulation bath. Khulbe et al. characterized those hollow fiber membranes by AFM [61]. Figure 4.35 shows the AFM images of four samples taken at four different positions along a hollow fiber spun at an air gap of 1 cm. For two samples (a and b), inner surfaces were observed at a scan size of 3 μm, and for the other two samples (c and d), three-dimensional images were taken at a scan size of 1 μm. From the pictures of samples a and b, the average, maximum, and minimum sizes of nodule aggregates were measured, and the results are shown in Table 4.9.

Table 4.9. Average, maximum, and minimum size (diameter) of nodular aggregates at the inner and outer surfaces of hollow fibers at air gaps 1 and 10 cm

Air gap (cm)	Figure	Average (nm)	Maximum (nm)	Minimum (nm)
Inside surface				
1	Fig. 4.35a	124	188	93
	Fig. 4.35b	123	143	95
10	Fig. 4.36a	140	170	117
	Fig. 4.36b	165	194	120
Outside surface				
1	Fig. 4.38a	124	186	109
	Fig. 4.38b	123	204	94
10	Fig. 4.39a	137	172	109
	Fig. 4.39b	144	148	94

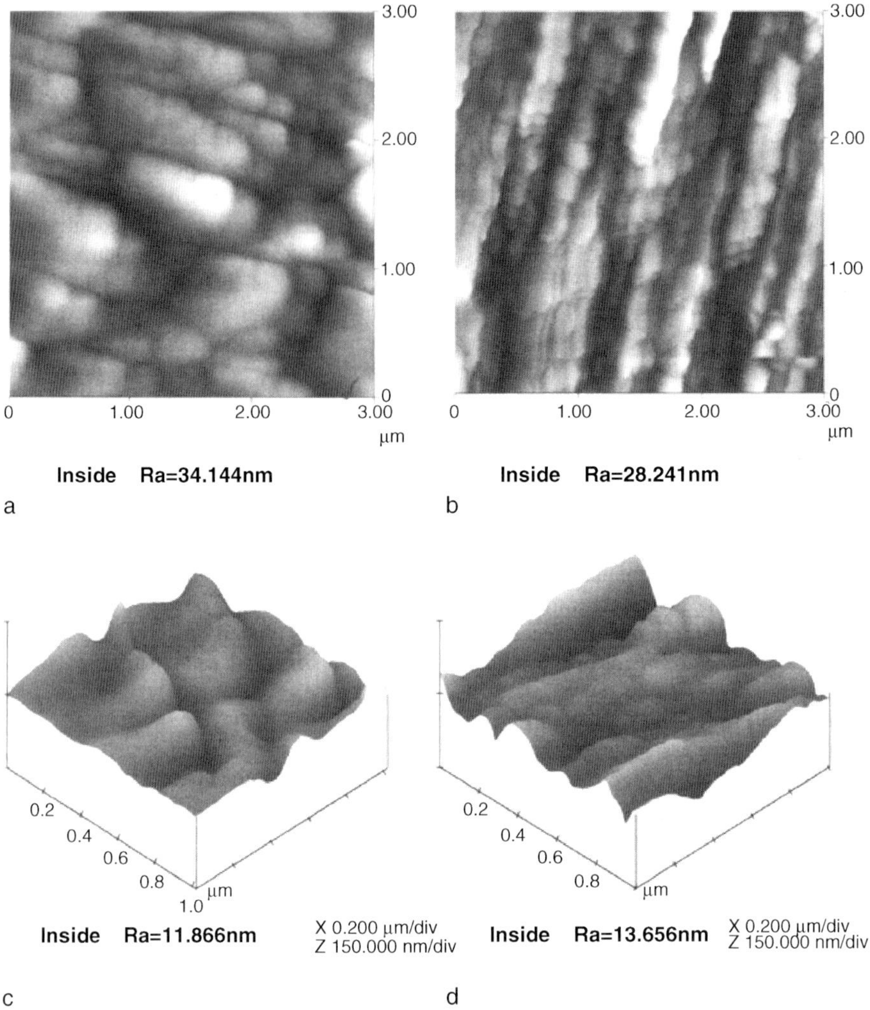

Fig. 4.35a–d. AFM images at four different sites of the surface of hollow fiber prepared at 1-cm air gap: **a** and **b** at scan 3 μm (top surface) and **c** and **d** at scan 1 μm (three-dimensional images). Reprinted from [61]. Copyright 2003, with kind permission from Elsevier

Although the AFM pictures are quite different, the sizes of nodule aggregates did not show any significant difference. In Fig. 4.35a, from three to six nodule aggregates are fused, while in Fig. 4.35b, several nodule aggregates are aligned in one row. It should be noted that the inner surface of the spin dope was in contact with a bore fluid that was running in a longitudinal direction at a speed higher than the spin dope, at the time when both extruded from the spinneret. (Eventually the velocity of the bore fluid and that of the dope solution would become equal.) A shear force worked on the inner surface of the nascent hollow fiber, causing alignment of nodular aggregates to

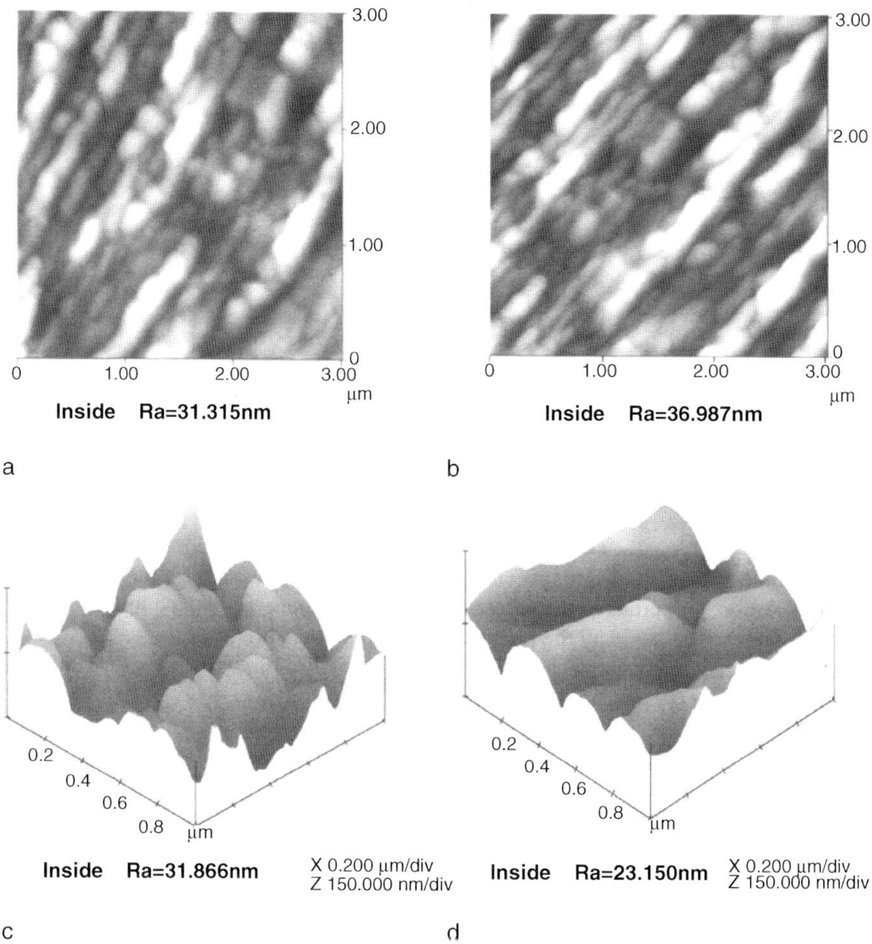

Fig. 4.36a–d. AFM images at four different sites of the surface of hollow fiber prepared at 10-cm air gap: **a** and **b** at scan 3 µm (top surface) and **c** and **d** at scan 1 µm (three-dimensional images). Reprinted from [61]. Copyright 2003, with kind permission from Elsevier

the direction of the bore fluid flow. The different images of samples a and b suggest that the alignment was not completed when the air gap was as short as 1 cm.

Figure 4.36a–d is for samples taken from inner surface of a hollow fiber spun at a 10-cm air gap. From the surface images a and b, the diameters of the nodule aggregates were obtained and the results shown in Table 4.9. Again, no significant difference in diameter was observed between samples a and b. However, Fig. 4.36a appears very similar to Fig. 4.36b.

Nodular aggregates are assembled to a number of string-like structures and aligned in one row. The similarity between Fig. 4.36a and b most likely means that the alignment of nodular aggregates under the shear force is completed when the air

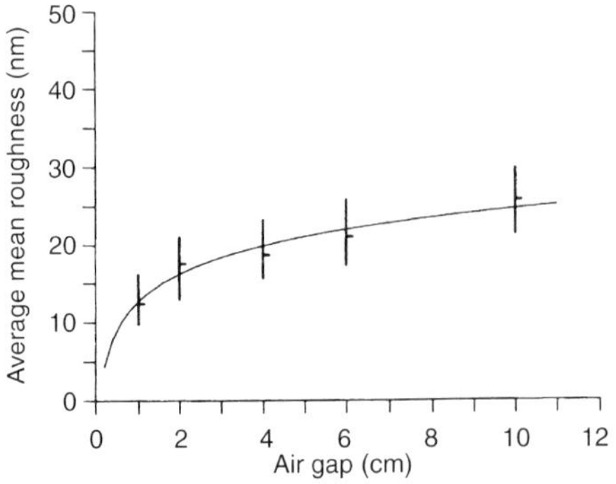

Fig. 4.37. Mean roughness of the inner surface of hollow fiber against air gap used for the preparation of hollow fiber membrane at a scan size of 1 μm. Reprinted from [61]. Copyright 2003, with kind permission from Elsevier

gap is as long as 10 cm. Interestingly, the elongation of the nodular aggregate itself was observed, as evidenced by the average length to width ratio of 1.25.

A comparison of Figs. 4.35 and 4.36 in their respective three-dimensional images (c and d) indicates that the roughness increases with the air gap. In fact, the roughness measured at several other different air gaps confirmed the above observation (Fig. 4.37).

Figures 4.38 and 4.39 show, respectively, the AFM images of the outer surfaces of the hollow fibers spun at the air gaps of 1 cm and 10 cm. Again, a and b are the surface images of two samples taken from two different sites along a hollow fiber, while c and d are the three-dimensional images of another two sites. The sizes (diameters) of the nodule aggregates summarized in Table 4.9 show that:

1. Comparing a and b, the aggregates sizes are not significantly different
2. Comparing the 1-cm and 10-cm air gaps, the aggregate sizes are not significantly different
3. Comparing the inner and outer surfaces, the aggregate sizes are not significantly different

By comparing the three-dimensional images of Figs. 4.38 (1-cm air gap) and 4.39 (10-cm air gap), it seems the outer surface becomes smoother as the air gap increases, unlike the inner surface. This was confirmed by measuring mean roughness parameters at several other air gaps, as shown in Fig. 4.40.

Comparing Figs. 4.37 and 4.40, it can be concluded that the outer surface is smoother than the inner surface. It is also interesting to note that, contrary to Chung et al.'s results, there was no alignment of nodular aggregates on the outer surface to any specific direction [58].

In another work, Khulbe et al. studied the surface morphology of poly(etherimide) hollow fiber membranes prepared by a dry–wet spinning method [62]. The air gap was changed in a broader range, from 10 to 90 cm. Note that the air gap in this study is much larger than the studies of Chung et al. [58] and Kapataidakis and Koops [59,60]. Figure 4.41 shows the fully developed alignment of nodular aggregates at the inner surface when the air gap is as long as 10 cm. The sizes of the nodular aggregates are summarized in Table 4.10 for both inner and outer surfaces. A decreasing

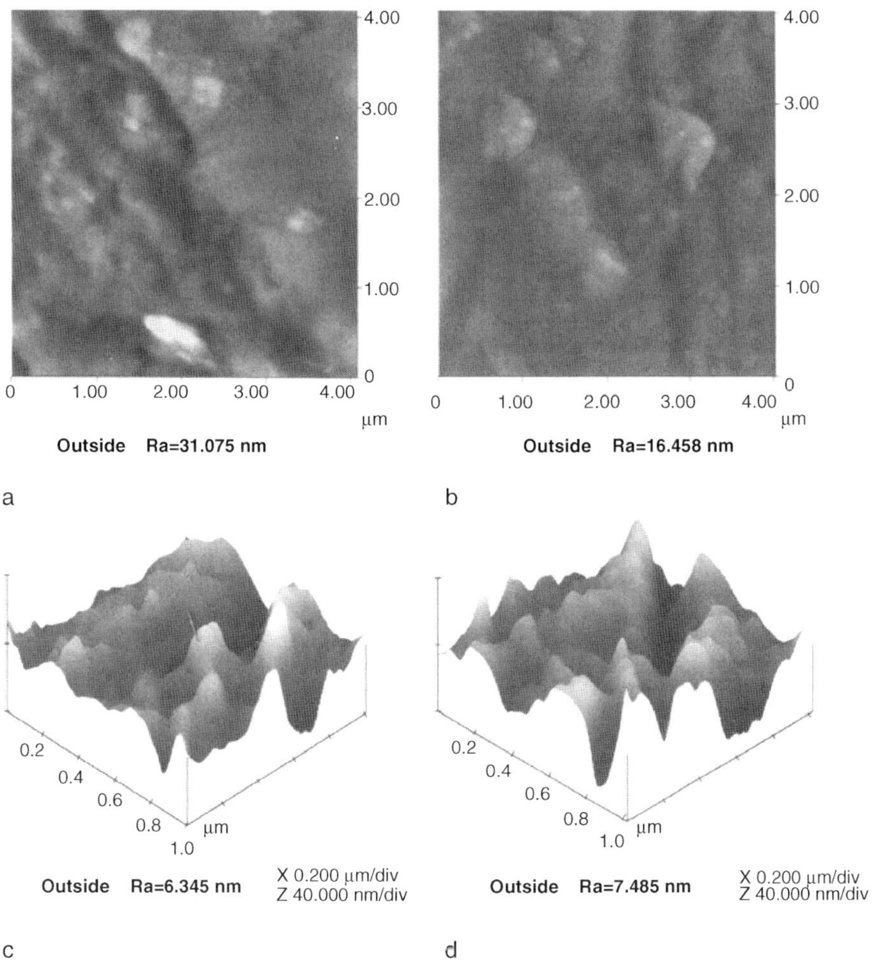

Fig. 4.38a–d. AFM images at four different sites of the outer surface of hollow fiber prepared at 1-cm air gap: **a** and **b** at scan 4 µm (top surface) and **c** and **d** at scan 1 µm (three-dimensional images). Reprinted from [61]. Copyright 2003, with kind permission from Elsevier

tendency in the average size of the nodular aggregates with an increase in the air gap is noticeable. For the change in the air gap from 10 to 90 cm, the decrease is about 10% at the inner surface, while it is about 40% at the outer surface. Since the nodular aggregates shrink as the hollow fiber travels a longer distance, the distance between the rows of the aligned nodular aggregates will increase. This is indeed observed in the second column of Table 4.10, where the distances between two rows of the nodular aggregates are shown.

It should also be noted that the molecular weight cutoff of the hollow fiber increased with an increase in air gap. Roughness parameters decreased as the air gap increased on both inner and outer surfaces.

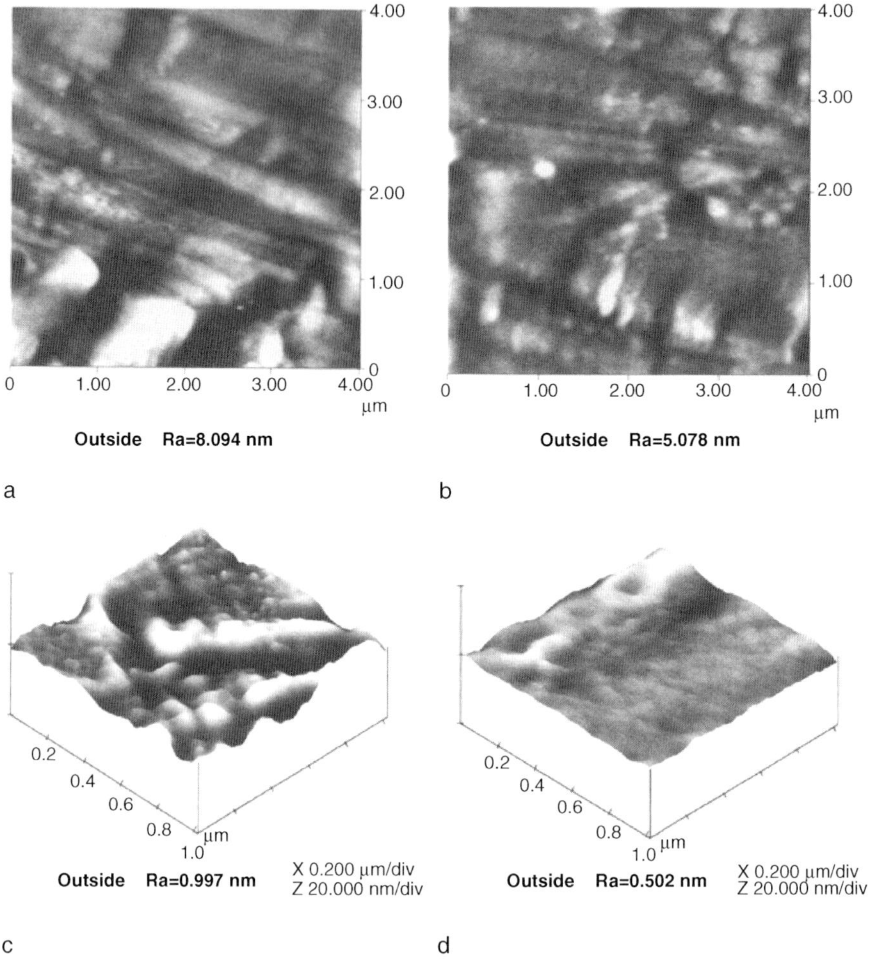

Fig. 4.39a–d. AFM images at four different sites of the surface of hollow fiber prepared at 10-cm air-gap: **a** and **b** at scan 4 µm (top surface) and **c** and **d** at scan 1 µm (three-dimensional images). Reprinted from [61]. Copyright 2003, with kind permission from Elsevier

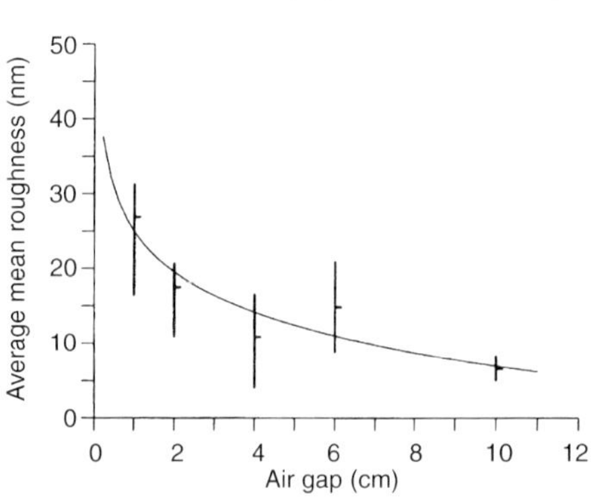

Fig. 4.40. Mean roughness of outer surface of the hollow fibers vs. air gap used for the preparation. Reprinted from [61]. Copyright 2003, with kind permission from Elsevier

Table 4.10. Average distance between the rows of nodular aggregates (inner surface) and average diameter of the nodular aggregates on the inner and outer surface of the hollow fibers prepared at different air gaps

Air gap (cm)	Average distance between the rows of nodular aggregates, inner surface (nm)	Average diameter of nodular aggregates (nm)	
		Inner surface	Outer surface
10	49.5 ± 3.6	100.0	50.8
30	48.7 ± 3.4	94.5	42.1
50	61.8 ± 4.8	95.3	34.0
70	59.2 ± 3.5	90.0	33.5
90	65.6 ± 4.9	91.6	30.8

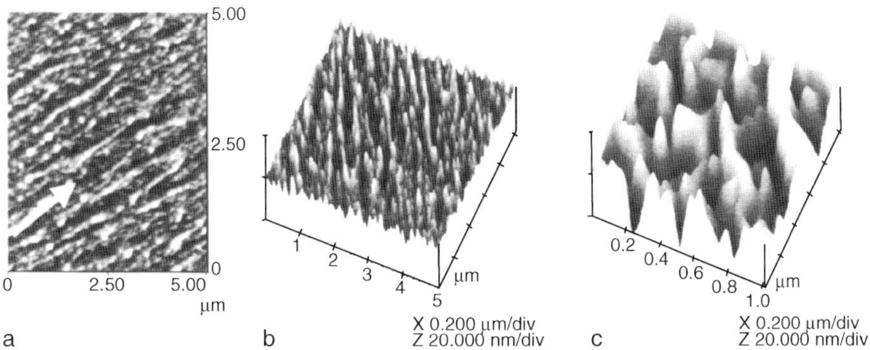

Fig. 4.41a–c. AFM images of inner surface of hollow fiber when the air gap is 10 cm. **a** At a 5-µm scan range (*arrow* shows the direction of bore fluid). **b** Three-dimensional image of **a**. **c** Three-dimensional image at 1 µm scan range. Reprinted from [62]. Copyright 2004, with kind permission from Elsevier

Feng et al. also studied the surface morphology of PEI hollow fiber membranes prepared by the dry–wet spinning method [63]. In their study, however, the bore liquid (water) flow rate was changed between 0.1 and 0.4 mL min^{-1}, while the air gap was maintained at 80 cm. Nodular aggregates were aligned to the direction of the bore fluid at the inner surface. However, they were only weakly aligned at the outer surface. The roughness parameter of the inner surface increased with an increase in the bore fluid flow rate, but the opposite was the case on the outer surface. Elongation of the nodular aggregate was observed, and the surface porosity increased as the bore liquid flow rate increased.

For hemodialysis membranes, biocompatibility is the primary requirement. It is known that surface properties such as surface roughness play important roles in determining membrane biocompatibility. It has also been reported that for a given material, smoother surfaces are more biocompatible [64]. Hence, the surfaces of three different commercial hollow fibers were studied by AFM to compare their roughness parameters. Figures 4.42 and 4.43 show AFM images of inner and outer surfaces, re-

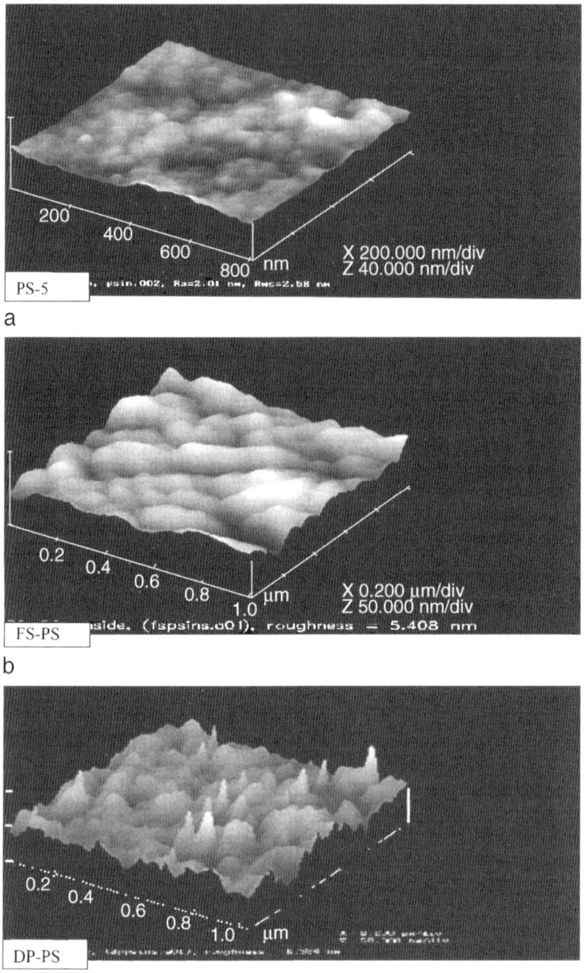

Fig. 4.42a–c. AFM images of the internal surfaces of typical PSf hollow fibers manufactured by **a** Baxter, **b** Fresenius, and **c** Membrana. Reprinted from [65], with kind permission from M. Asmanrafat

Table 4.11. Data on nodule size and roughness parameter at the internal and external surfaces of three PSf hemodialysis hollow fiber membranes manufactured by Baxter, Fresenius, and Membrana

Hollow fiber	Manufacturer	Internal surface Mean nodule size (nm)	Internal surface R_a (nm)	External surface Mean nodule size (nm)	External surface R_a (nm)
PS-5	Baxter	2.5 ± 1.6	2.5 ± 0.5	8.2 ± 1.6	10.4 ± 2.5
DP-PS	Membrana	14 ± 1.76	6.4 ± 0.7	27.1 ± 1.4	33.5 ± 9.6
FS-PS	Fresenius	25 ± 15	5.4 ± 0.8	40.7 ± 1.5	98.0 ± 15.3

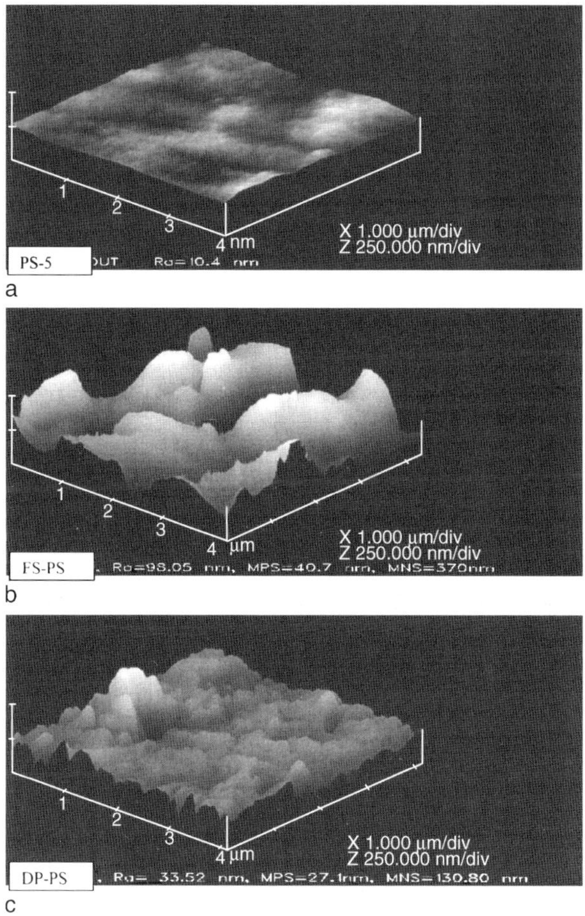

Fig. 4.43a–c. AFM images of the external surfaces of typical PSf hollow fibers manufactured by **a** Baxter, **b** Fresenius, and **c** Membrana. Reprinted from [65], with kind permission from M. Asmanrafat

spectively, of the polysulfone hollow fibers manufactured by Baxter, Fresenius, and Membrana [65]. In all three hollow fibers, the nodules are aligned in rows on both surfaces. However, the alignment is more visible on the inner surface. The average nodule sizes and the roughness parameters of membranes manufactured by Baxter, Fresenius, and Membrana are summarized in Table 4.11. Baxter's membrane was found to possess the smoothest surface among those membranes. From Table 4.11, it can be assumed that the PSf hollow fibers manufactured by Baxter Inc. have a better biocompatibility compared to Fresenius and Membrana.

Barzin et al. characterized poly(ether sulfone) (PES) hollow fibers for hemodialysis by both ultrafiltration experiments and AFM [66]. Hollow fibers were fabricated from poly(ether sulfone) (Ultrason E6020; M_w 58 000; flakes from BASF Co.) by the dry–wet spinning technique at the conditions similar to those of Gholami et al. [67].

Gholami et al., however, used different PES (Victrex 4100 P, Imperial Chemical Industries) and studied the effect of heat treatment at different temperatures (120 – 210 °C) for the development of ultrafiltration membranes. Going back to Barzin et al.'s work, poly(vinylpyrrolidone) was added to the spinning dope. Two solutions with different PES/PVP weight ratios (18/6 and 18/3) in dimethylacetamide solvent were used as spin dopes. After the spinning, hollow fibers were heated either in water (95 °C for 30 min) or in air (150 °C for 5 min). Two-dimensional images of the inner surfaces of the studied hollow fibers showed that nodules were in rows and aligned in the direction of bore fluid (water) flow.

Hayama et al. studied the inner and outer surface of a hollow fiber dialysis membrane (polysulfone, APS-150, Asahi Medical, Japan) by using a silicon single-crystal probe in AFM. From the AFM images, it was observed that the nodules/nodular aggregates were in rows on the inner surface of the hollow fibers [68]. However, no such rows of nodules/nodular aggregates were observed on the outer surface. It seems that nodules and nodular aggregates are generally better aligned in the inner surface than outer surface.

4.4 Effects of Membrane Preparation and Posttreatment Parameters on the Nodular Size

Usually, dense flat sheet membranes are prepared by evaporating the solvent at room temperature. Khulbe et al. prepared dense poly(phenylene oxide) membranes from a trichloroethylene (TCE) solution by evaporating the solvent at 22, 4, and −10 °C and reported that the morphology of the surface depended on the solvent evaporation temperature [69]. The average diameter of the nodule and the mean roughness of the membrane's top surface are summarized in Table 4.12. An extremely large nodule diameter (most likely this is the diameter of the supernodular aggregates) and mean roughness were observed when the temperature was as low as −10 °C.

Khulbe et al. prepared dense PPO membranes of different thicknesses by pouring different amounts of 0.25 wt.% PPO solution in TCE solvent in an aluminum ring and drying the solvent completely [70]. Then the top surface morphology was studied by AFM technique. The nodule/nodular aggregate size and the roughness parameter were then correlated to Marangoni and Rayleigh numbers. Figure 4.44 shows the TM-AFM image of the top surface of a 2-µm-thick membrane, of which the average nodule/nodular aggregate size is 46 nm. Similar AFM images were ob-

Table 4.12. Summary of the average nodule diameter and mean roughness parameters for top surfaces of PPO membranes prepared at different temperatures

Evaporation temperature °C	Average nodule diameter (nm)	Mean roughness [a] (nm)
22	49.0	0.17
4	33.3	0.22
−10	374.5	8.14

[a] Scan size 1 µm (for roughness measurement)

tained for other membranes when the thickness was increased up to 9 µm. The size of the nodules decreased with an increase in the thickness of the membrane until the thickness became 9 µm. However, at a thickness of 10 µm, some *vortexes* were observed [71]. On further increase in film thickness, these vortexes were converted to supernodular aggregates. Figure 4.45 shows the TM-AFM image of an 11-µm membrane, which contains supernodular aggregates. The average size of the supernodular aggregates is 890 nm, and nodules are observed inside the supernodular aggregates. The sizes of both nodules and supernodular aggregates started to increase from this thickness. Thus, a minimum in the nodule (supernodular aggregate) size appeared between 9 and 11 µm. Figure 4.46 shows the effect of the membrane thickness on the sizes of nodules/supernodular aggregates. The figure also includes the data obtained from the membranes cast from 0.5 wt.% of PPO in TCE solvent. Figure 4.47 shows the effect of the membrane thickness on the mean roughness of the surface. The minimum mean roughness appears at a membrane thickness between 9 and 11 µm when membranes are cast from the 0.25 wt.% PPO solution (Fig. 4.47, line *I*). Figure 4.47

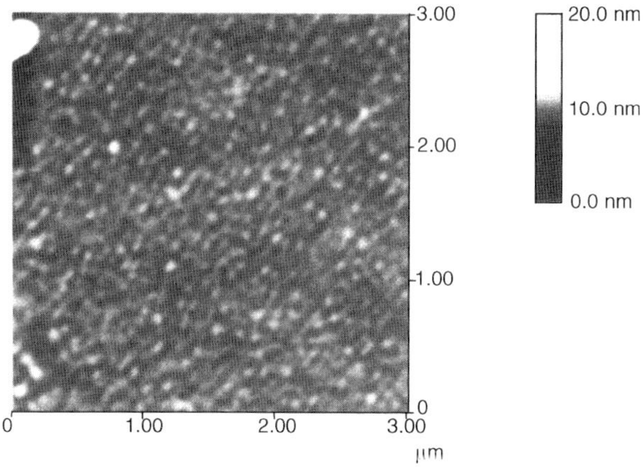

Fig. 4.44. AFM image of the top surface of a 2-µm-thick PPO membrane prepared from 0.25 wt.% PPO in TCE. Bright spots are the nodules. Reprinted from [70]. Copyright 1998, with kind permission from Elsevier

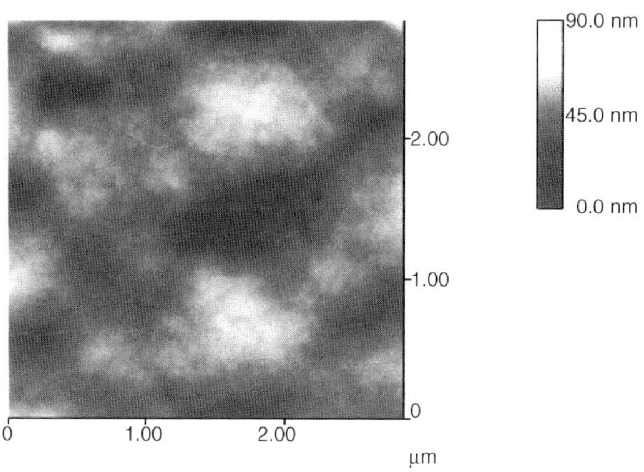

Fig. 4.45. AFM image of the top surface of a 11-µm-thick PPO membrane prepared from 0.5 wt.% PPO in TCE. Bright spots are the nodules. Reprinted from [70]. Copyright 1998, with kind permission from Elsevier

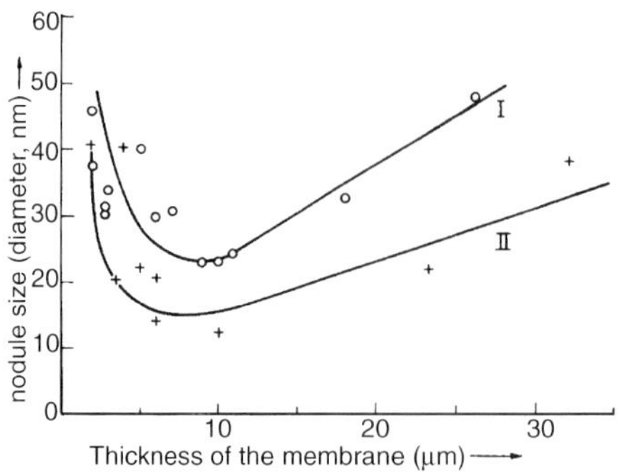

Fig. 4.46. Effect of the thickness of the membrane on the sizes of the nodules (top surface). *I* indicates the membrane prepared from 0.25 wt.% PPO in TCE. *II* indicates the membrane prepared from 0.5 wt.% PPO in TCE. Reprinted from [70]. Copyright 1998, with kind permission from Elsevier

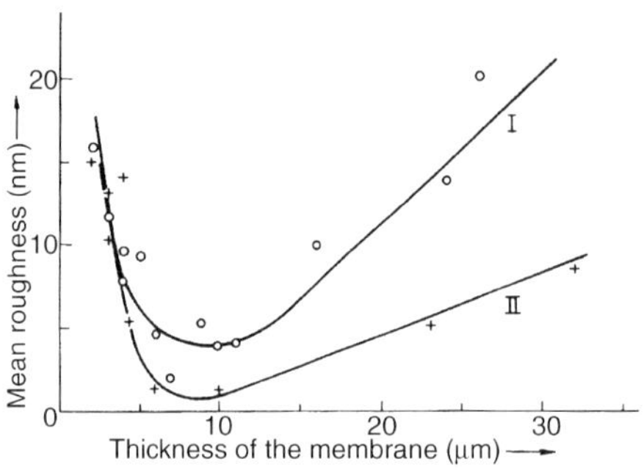

Fig. 4.47. Effect of the thickness of the membrane on the mean roughness (top surface): *I* indicates the membrane prepared from 0.25 wt.% PPO in TCE. *II* indicates the membrane prepared from 0.5 wt.% PPO in TCE. Reprinted from [70]. Copyright 1998, with kind permission from Elsevier

line *II* shows that the minimum appears also when the membrane is prepared from 0.5 wt.% PPO solution; however, the position of the minimum shifts toward the lower end of the thickness.

The effect of the membrane thickness on the surface morphology can be explained by considering the initial instability of the casting film, which is caused by surface tension gradient and the density gradient. The instability of liquid films when subjected to density and/or surface tension gradients causes convection flow-forming vortex cells known as Bernard cells. The Marangoni (*Ma*) and Rayleigh (*Ra*) numbers are often used to determine the conditions for the onset of the formation of convection cells caused by the surface tension gradient and density gradient, respectively. *Ma* and *Ra* can be calculated in the following way:

The temperature difference between the top and the bottom of the cast film, ΔT, is calculated by two equations (Eqs. 4.1 and 4.2):

$$N_{\mathrm{A}} = \frac{D_{\mathrm{AB}} P}{RT(z_2 - z_1) P_{B,lm}} (P_{\mathrm{A1}} - P_{\mathrm{A2}}) \quad (4.1)$$

where N_{A} is the evaporation rate per unit area (kg – mol s^{-1} m^{-2}), D_{AB} is the diffusivity of TCE solvent (A) in air (B) (m^2 s^{-1}), P is the atmospheric pressure (kPa), P_{A1} is the partial vapor pressure of the solvent (kPa) at the air/film interface (which equals the saturation vapor pressure of the solvent), P_{A2} is the partial vapor pressure of the solvent at the end of the stagnant air film (kPa) (in this case, set equal to zero.), $P_{B,lm}$ is the logarithmic mean of the partial pressure of air (which equals $(P_{\mathrm{A1}} -$

$P_{A2})/\ln[(P - P_{A2})/(P - P_{A1})])$, R is the gas constant (8.314 m^3 kPa kg – mol^{-1} K^{-1}), T the temperature (K), $(z_2 - z_1)$ is the thickness of the stagnant air film (m) (the height of the aluminum ring which was used as a mold for the preparation of the membrane, 0.005 m) and

$$\Delta T = \frac{N_A h \lambda}{k} \tag{4.2}$$

where h is the thickness of the cast film (m), λ is the heat of vaporization of the solvent (kJ kg – mol^{-1}), and k is the thermal conductivity of the solvent (kJ s^{-1} m^{-1} K^{-1}).

It should be noted that the cast solution is considered to be very dilute in this calculation. Equation 4.1 is based on the steady-state diffusion of solvent vapor through stagnant air [72], and Eq. 4.2 is based on the heat balance.

The Marangoni number based on the surface tension is given as

$$Ma = \frac{\mathrm{d}\gamma}{\mathrm{d}T} \frac{\Delta T h C_p}{\mu k} \tag{4.3}$$

where $\mathrm{d}\gamma/\mathrm{d}T$ is the surface tension gradient (N m^{-1} K^{-1}), C_p is the heat capacity (J kg^{-1} K^{-1}), ρ is the density (kg m^{-3}), and μ is the viscosity (Pa s).

The Rayleigh number based on the density gradient is given as

$$Ra = \frac{\alpha G \Delta T h^3 \rho^2 C_p}{\mu k} \tag{4.4}$$

Where α is the thermal volume expansion coefficient (°C) and G is the gravity constant (9.80 m s^{-2}).

For the top boundary insulating air, such as in the case of the casting film on the glass, the critical Marangoni number (*Mac*) and Rayleigh number (*Rac*) for the onset of convection of the free surface liquid due to the temperature gradient were suggested as 79.6 and 669, respectively [73]. *Ma* and *Ra* numbers for the casting films were calculated to determine if surface tension and gravity-driven convection takes place, which may cause instability of the casting film in the beginning of film formation. Table 4.13 summarizes the results for the membranes of different thicknesses. The casting film thickness when the film was wet was evaluated from the casting solution composition (PPO 0.25 wt.%) as was the final membrane thickness when the membrane was dry. The viscosity of the casting solution was considered to be that of the solvent since the polymer concentration was low. Due to the fast evaporation of the solvent, the temperature difference, ΔT, at the beginning of the membrane casting was used in Eqs. 4.3 and 4.4. Variation of temperature and viscosity of the casting solution with time was not considered, but the use of the initial temperature difference to estimate the *Ra* and *Ma* numbers will provide a general trend of membrane surface morphology by the onset instability of the casting solution.

The sharp increase in roughness of membranes less than 9 μm thick could be due to the initial instability of the casting film caused mainly by surface tension gradients, since the Rayleigh numbers of the membranes are much smaller, in most cases, than

Table 4.13. Estimated Marangoni and Rayleigh numbers for different film thicknesses of PPO casting solution (0.25 wt.%)

Membrane thickness (dry) (μm)	Casting film thickness (wet) (μm)	Ma	Ra	*Ma*/*Mac*	*Ra*/*Rac*
0.5	69	30.4	0.017	0.35	2.4×10^{-5}
1	138	127.1	0.27	1.4	3.9×10^{-4}
2	276	486.7	4.3	5.6	0.006
3	414	1095.2	22.0	12.6	0.03
4	552	1947.0	69.4	22.4	0.10
5	690	3042.2	169.4	35.0	0.24
6	828	4380.7	351.3	50.5	0.50
7	966	5962.7	650.9	68.7	0.93
8	1104	7788.0	1110.3	89.7	1.59
9	1242	9856.6	1778.5	113.5	2.54
11	1518	14 724.1	3968.8	169.6	5.68
24	3312	70 091.7	99 936.4	807.5	128.7
26	3864	95 402.60	66 618.2	099.1	238.4

the critical *Ra* as shown in Table 4.13. For casting films with a thickness less than 1000 μm (which correspond to a membrane thickness of 7 μm), the surface tension gradient tends to be the main source of convection cells. As the initial thickness of the casting solution lessens, the surface tension gradient and uneven evaporation of solvent tend to generate instability, which is so fast and vigorous that a rougher and irregular surface is quickly formed. This phenomenon is called initial instability as compared to the formation of regular hexagonal cells due to the convective flow. Temperature differences required for the initial instability were found to be much smaller for thinner layers. In other words, when the membrane becomes thinner, the critical Marangoni number becomes smaller than 79.6, a value that was used to calculate (*Ma*/*Mac*) in Table 4.13. Scriven and Sterling [74] also suggested that smaller onset temperature gradients for thinner layers are accompanied by increased wavelengths or cell size. Therefore, the rougher membrane surfaces were obtained for thinner membranes of 0.25 wt.% PPO by weight solutions, with thicknesses less than 9 μm. The surface roughness of the membrane with a thickness greater than 11 μm was affected by the surface tension gradient as well as the density gradient. The *Ra* and *Ma* numbers are much larger than their critical values, as indicated in Table 4.13.

Koschmieder and Biggerstaff [75] also observed that for liquid films with thicknesses more than 2 mm and high *Ma* and *Ra* values, there was no initial instability that would result in the formation of unstable cells, but uniform regular hexagonal cells were formed. The cell size became larger as the *Ma* and *Ra* numbers increased.

The effect of viscosity on the formation of the cells is seen by the change in the polymer concentration in the cast polymer solution. The increase in viscosity tends to resist the formation of the convective flow. The membrane becomes, therefore, smoother for the polymer solution of higher polymer concentration. For example, membranes made from 0.5 wt.% PPO solution have smoother surfaces than mem-

Table 4.14. Roughness parameter for membrane surfaces of three samples

Sample	R_a (nm)
M1	3.1
M2	1.7
M3	9.9

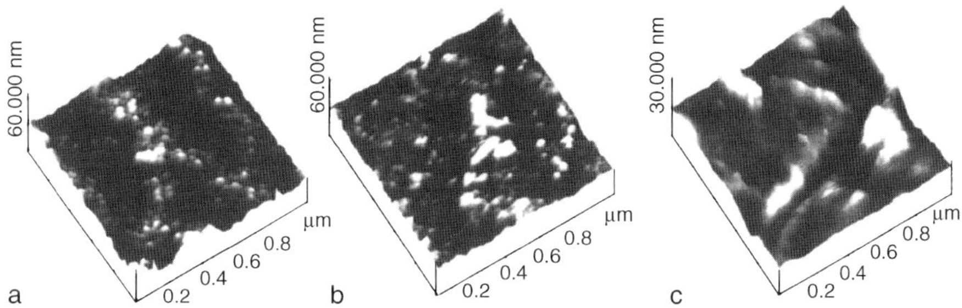

Fig. 4.48a–c. Surface plot of the image by TM-AFM of the **a** M1, **b** M2, and **c** M3 samples. Reprinted from [76]. Copyright 1998, with kind permission from Wiley

branes made from 0.25 wt.% solution. The thickness that gave minimum roughness shifted toward the left as shown in Fig. 4.47, since the range of the initial instability decreased. These phenomena could also be explained by the *alligator* skin layer effect as described by Kesting [71]. This effect is more frequently encountered when dilute solutions and complete evaporation techniques are employed.

Kim et al. discussed surface structure and the phase separation mechanism of polysulfone membranes by AFM [43]. A membrane formed by immersion in a pure water coagulation bath showed a nodular structure with a nodule size of about 25 nm, which was believed to be the result of spinodal decomposition. A membrane formed by immersion in a coagulation bath mixture (water/NMP 20/80 by weight) had the porous structure with a mean radius of 146 nm, which was the result of nucleation and growth of the polymer-poor phase.

Lehmani et al. studied the surface morphology of Nafion 117 membranes [76] by TM-AFM. Three different samples were prepared. The sample denoted M1 was dried under vacuum at 80 °C. The samples M2 and M3 were first dried, and, before AFM imaging, a drop of deionized water and a drop of tributyl phosphate (TBP), respectively, were placed on the surfaces of the membranes in order to determine the influence of the swelling properties of the membranes. Figure 4.48 shows the three-dimensional AFM images of those membranes. Figure 4.48a of the M1 membrane shows spherical grains with a mean diameter of 11 nm. Table 4.14 shows the roughness parameters for the membrane surfaces.

Table 4.14 and Fig. 4.48 illustrate that M3 is quite different from M1 and M2. It is known that the volume increase of the membrane takes place upon absorption of TBP (~ 360%) [77]. Thus, the large change noticed in M3 is due to the swelling of the membrane surface. This occurred because the polar site of the membrane was hydrated.

Soresi et al. [78] studied the surface morphology of a dense homopolymer poly(vinylidene fluoride) film (thickness 80 μm, Goodfellow) and a dense and transparent film of the copolymer poly(vinyldenefluoride-hexafluoropropylene), P(VDF-5% HFP). Both membranes were prepared by the conventional casting technique. These membranes were developed as proton exchange membranes for fuel cells. In addition to these membranes, they also studied two porous PVDF homopolymer membranes of different pore sizes. Both PVDF and P(VDF-HFP) membranes were grafted by styrene followed by sulfonation. The authors were interested in studying the influence of the base matrix, in terms of polymer nature and morphology, on the functionalization process and physical chemistry of the sulfonated membranes. Figure 4.49a and b shows the AFM images of the porous PVDF membrane (pore dimension 100 nm) and dense P(VDF-HFP) membrane, respectively. Figure 4.49c and d shows AFM images after the grafting process on the above PVDF and P(VDF-HFP) membranes, respectively. The PVDF membrane (Fig. 4.49a) shows a globular structure, in which pores with dimensions of around 100 nm are homogeneously distributed. The polymer domains, ranging from 500 nm to 1 μm, have parallel orientations, which is related to the film processing. After grafting (Fig. 4.49c), the topography of the matrix changed significantly. The parallel orientations of the domains were compromised by a disordered rearrangement induced by styrene unit grafting on the PVDF chains. The polymer globular dimensions consequently increased up to 3 μm as did the surface roughness. Similar effects on the surface morphology due to the grafting process can be detected in the dense copolymer P(VDF-HFP) membrane (Fig. 4.49b and d).

Zhang et al. prepared cellulose membranes from a mixture of cellulose, water, and *N*-methylmorpholine-*N*-oxide (NMMO) under different preparation conditions [79]. These membranes showed different UF permeation properties, which could be explained very well with AFM investigations. Increasing the temperature and NMMO concentration of the coagulation bath led to higher values for the roughness parameters, larger pore sizes, etc. However, MWCO was not measured.

Oh et al. studied PA composite membranes prepared by the conventional interfacial polymerization of PA active layers on the surface of various microporous polyacrylonitrile (PAN) supports [80]. The PAN supports were prepared by using PAN/NMP solution with various compositions (10/90, 15/85, and 20/80 wt.%). The PAN supports were further modified with NaOH solutions of different concentrations for 1 h at 40 °C to form –COOH groups on their surfaces [81]. Figure 4.50 shows AFM photographs of PAN membranes treated with different NaOH concentrations after their formation from a 15 wt.% PAN solution. Figure 4.50 indicates visually the difference in the surface morphology between those membranes. The surface became smoother as the concentration of NaOH increased.

Figure 4.14 shows AFM photographs of the PA composite membranes prepared by using modified PAN supports that were prepared from different PAN concentrations, i.e., 10, 15 and 20 wt.%. The surface roughness of each was different, and the

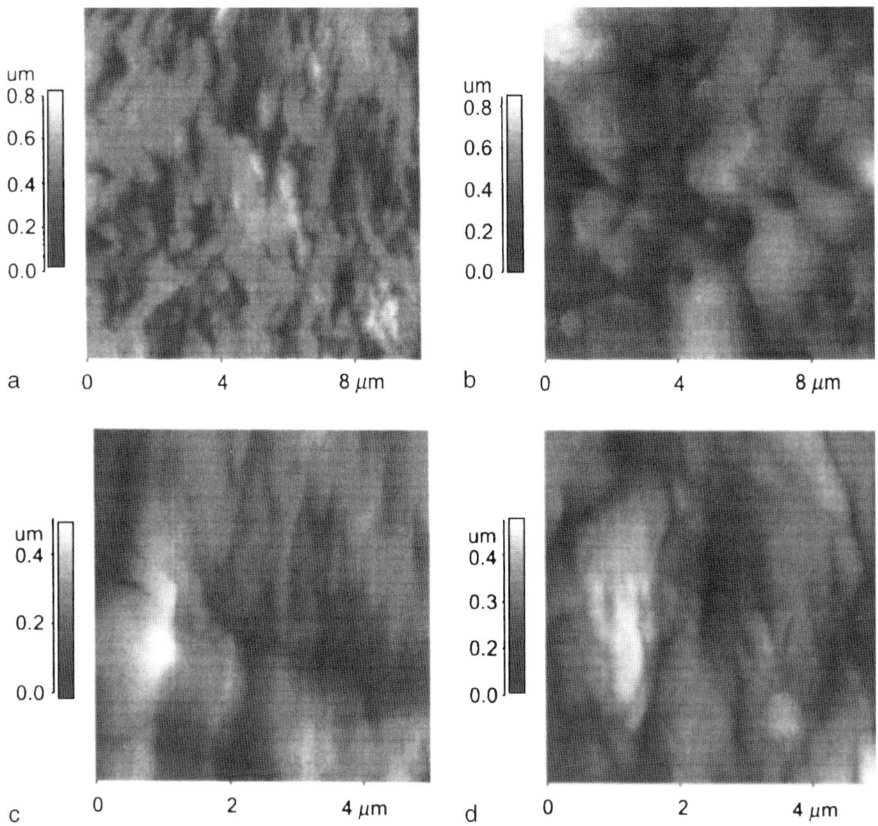

Fig. 4.49a–d. AFM images of the PVDF porous membrane (pore dimension 100 nm) and those of the dense P(VDF-HFP) before (**a** and **b**) and after (**c** and **d**) the grafting process. Reprinted from [78]. Copyright 2004, with kind permission from Elsevier

one prepared from higher PAN concentration showed a smoother surface. In general, the surfaces of the polymeric membranes made from the solutions with higher polymeric concentrations by the phase inversion method are smoother in comparison to the ones made from lower concentrations of polymer solution. Figure 4.50 indicates that the surface roughness of the resulting PA composite membranes depends on the roughness of the supports. From these results, the surface morphology or roughness of the PA composite membranes is in close relationship with that of the supports.

Stamatialis et al. [82] studied by AFM the structures of dense and integrally skinned cellulose acetate (CA) and cellulose acetate butyrate (CAB) membranes prepared by the phase inversion technique under different casting conditions. Figure 4.51a and b shows the AFM images of the top (casting solution and air interface) and bottom (casting solution and glass plate interface), respectively, of the dense CA membrane. Both surfaces show a relatively uniform structure, despite the fact that the R_a parameter of the bottom surface is higher (0.93 nm) than the top surface (0.4 nm).

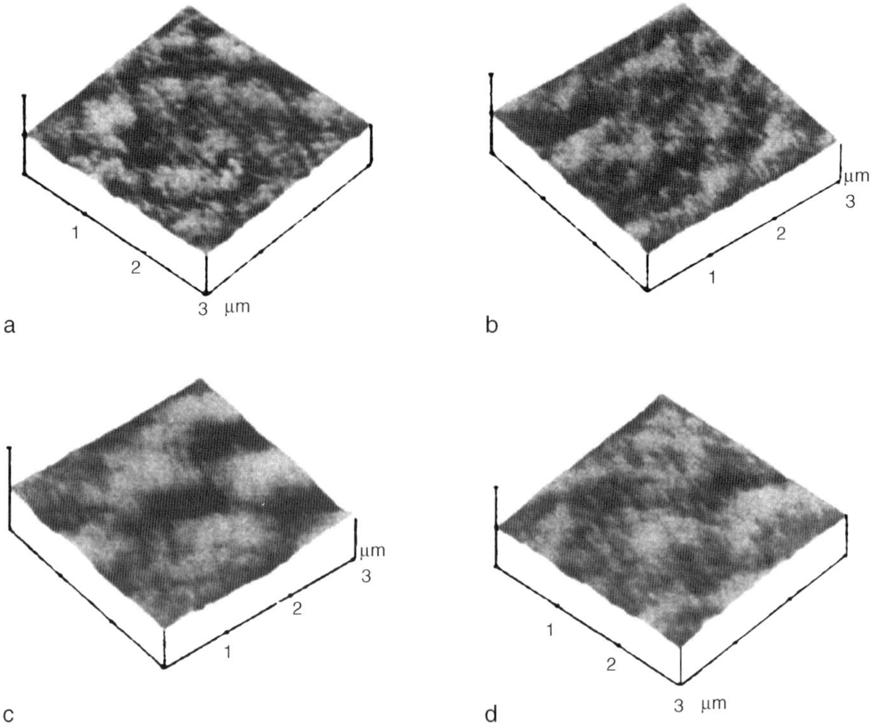

Fig. 4.50a–d. AFM photographs of PAN membranes treated for 1 h with different NaOH concentrations at **a** 0.1, **b** 0.5, **c** 1, and **d** 2 M, after their formation from a 15% PAN solution. Reprinted from [81]. Copyright 2001, with kind permission from Wiley

Visually, the size of the nodule or nodule aggregates on the bottom surface is bigger than those on the top surface. The top and the bottom surfaces of the dense CA membrane were reported to be quite uniform in comparison with the corresponding surfaces of the asymmetric CA and CAB membranes. However, the study did not give any AFM data for the dense CAB membrane. The roughness parameter of the active layer of the asymmetric CA membrane tends to decrease as the solvent evaporation time increases. Thus, asymmetric membranes prepared under different casting conditions showed a wide range of NF/RO permeation characteristics. These characteristics depended on the surface topographies of the active layer.

Alsari et al. studied the effect of sodium dodecyl sulfate (SDS) solutions as gelation media on the surface properties of poly(ether sulfone) membranes [83]. The temperature of the gelation media was either 4 or 20 °C. The concentration of SDS was changed from 0 to 3.0 $g\,L^{-1}$ at 4 °C and from 0 to 1.6 $g\,L^{-1}$ at 20 °C. From the analysis of AFM images, it was observed that the presence of SDS in the gelation media had an effect on the membrane surface morphology as well as on the membrane performance. Figure 4.52 shows the effect of SDS concentration in the gelation bath on the molecular weight cutoff values of the PES membranes. A similar trend

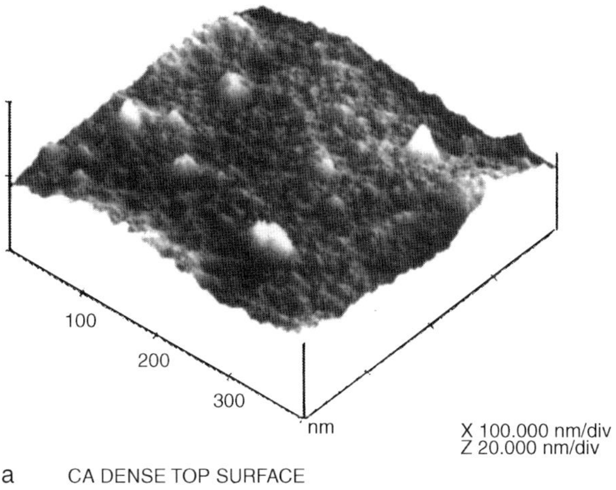

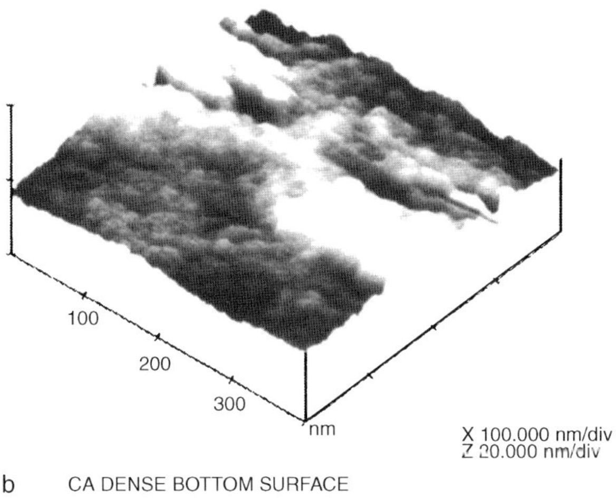

Fig. 4.51a,b. AFM images of the dense symmetric CA membrane at **a** top and **b** bottom surfaces. Reprinted from [82]. Copyright 1999, with kind permission from Elsevier

was observed for the roughness parameter versus SDS concentration in the gelation bath. However, data was not given for the size of the nodules or nodule aggregates at the surface. Comparing the two temperatures at which the membranes were gelled, the effect of SDS concentration on the membrane morphology and performance was more pronounced at 4 °C than at 20 °C. The pore size increased as the surface roughness increased.

Xu and Coleman studied 6FDA (2,2′-bis(3,4-dicarboxyphenyl)hexafluoropropane dianhydride)–pMDA (pyromellitic dianhydride) polyimide films irradiated by an ion beam [84]. A beam of 140 keV N^+ions with a low-current density was used, and three irradiation fluences (2×10^{14} cm^{-2}, 1×10^{15} cm^{-2}, and 5×10^{15} cm^{-2}) were chosen. It was reported that even a small dose altered the microstructure of the surface layer. The AFM analysis of those films showed that low-fluence irradiation induced microvoids in the surface layer of the polymer, and high-fluence irradiation resulted

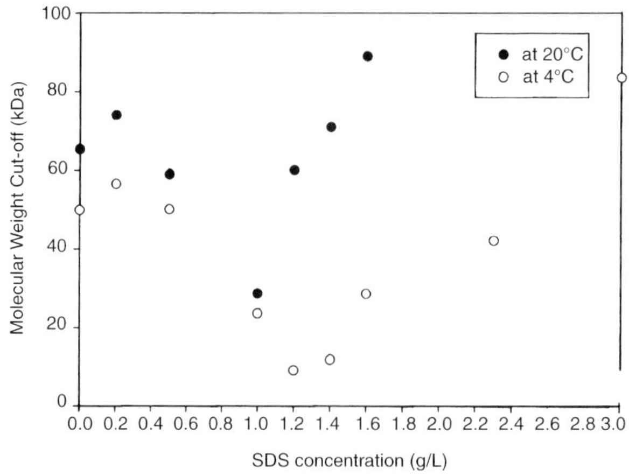

Fig. 4.52. Effect of SDS concentration in the gelation bath on the MWCO of a PES membrane. Reprinted from [83]. Copyright 2001, with kind permission from Elsevier

in a large number of small microvoids in the surface. All of these results agree well with the ion-beam irradiation effects on the iodine diffusion and gas permeation properties of the polyimide. However they do not give data for nodule or nodule aggregates on the surface.

Broadhead and Tresco studied the effects of fabrication conditions on the structures and performances of membranes formed from poly(acrylonitrile-vinylchloride) (PAN-PVC) by using the phase inversion process [85]. They reported the relationship of the fine-surface structure of PAN-PVC membranes to the membrane performance and membrane fabrication method. The fine-surface structure of nodular elements and the size of these elements could be altered by changing the precipitation conditions. Membranes were prepared at 22 °C on 55 mm diameter polished silicon wafers by spinning at 1500 rpm for 20 s with a spin coater [86]. The film was immediately precipitated in one of the four different precipitation media. The first three media consisted of deionized water at 4, 22, and 54 °C. These membranes were referred to as "Type 1", "Type 2", and "Type 3", respectively. The fourth medium was a 50/50 mixture of deionized water and *N*,*N*-dimethylformamide (DMF) at 54 °C and coded as "Type 4". Figure 4.53 shows the histograms of the nodule size distributions observed at the skinned surface of the membranes made under four different precipitation conditions. The sizes of these nodular elements became smaller and more uniform with milder precipitation conditions, which supports the theory that nodules are formed through spinodal decomposition under these conditions. In addition, the size of these nodules could be related to water permeability. Hence, water transport occurred through the interstitial spaces where the pores could be situated.

4.5 Summary

Table 4.15 summarizes the applications, pore sizes, nodule/nodular aggregate sizes, and roughness (surface) parameters for different polymeric membranes that are covered in this chapter. The following conclusions can be drawn from this table:

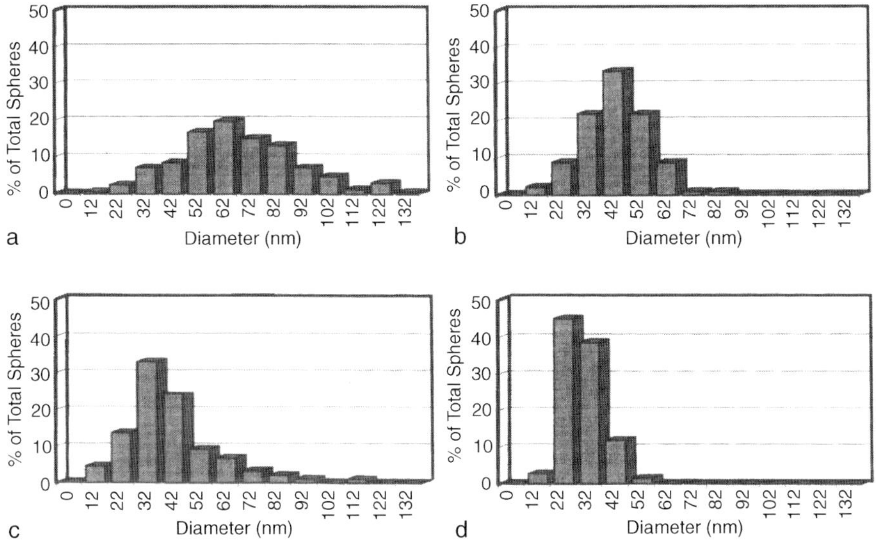

Fig. 4.53a–d. Histograms of the nodule size distribution observed at the skinned surface of membranes made under different precipitation conditions: **a** Type 1, **b** Type 2, **c** Type 3, and **d** Type 4. Please note the shift in size and dispersiveness of the nodular elements. Reprinted from [85]. Copyright 1998, with kind permission from Elsevier

1. There seems to be no relationship between the membrane application and the nodule size.
2. There seems to be some relationship between the membrane application and the roughness parameter.
3. For gas separation membranes, roughness parameters cover a broad range of 0.134–9.6 nm. The lower ends are in most cases less than 1 nm. Exceptions are PSI-PI (31–36 nm) and HMDSO (14 nm). It should be noted that the PSI-PI membrane is before silicone coating, and the HMDSO membrane is after plasma polymerization. Both membranes show no gas separation.
4. Integrally skinned asymmetric reverse osmosis and nanofiltration membranes also feature small roughness parameters, ranging from 0.84 to 5.14 nm.
5. Roughness parameters of composite RO and nanofiltrtion membranes are an order of magnitude higher than the integrally skinned asymmetric RO membranes, ranging from 10 to 82 nm. This reflects the roughness parameters of porous substrate membranes, which are in most cases ultrafiltration membranes.
6. Roughness parameters of UF membranes range from 1.21 to 66 nm. They are between RO and microfiltration membranes.
7. Roughness parameters of MF membranes are the largest, ranging from 67 to 96 nm.

Table 4.15. A summary of application, pore sizes, nodule/nodular aggregate sizes, and roughness

Material	Application	Pore size [a] or MWCO (Da)	Nodule size [a]	Roughness (nm)	Reference
PPO	Gas	Dense	44–60	1–3.5	[16]
PPO	Gas	Dense	43–69	1.0–3.46	[24]
PES-PI (hf)	Gas	Dense	124–165	31–36	[61]
HMDSO	Gas	–	20	14	[48]
CA	Gas	Dense	–	0.4–0.93	[82]
CA asymmetric	RO and NF			0.91–5.14	
Cellulose acetate butyrate	RO and NF			0.87–0.94	
6FDA-APPS	Gas	Dense	–	0.89	[15]
6FDA-APPS	Gas	Asymmetric	–	1.5–6	[15]
Poly(3-(2-acetoxyethyl)thiophene)	Gas	–	–	51–67	[87]
CA	Gas	Dense	95–600	0.437–9.6	[88]
Polyimide	Gas	Dense	–	0.578	[84]
Ion-implanted 6FDA-pMDA polyimide	Gas	Composite	–	0.134–0.366	[84]
PPO	Gas	Dense	15–20	0.6–1.3	[88]
PPO	Gas	Composite	45	4.2	[88]
Acrylamide/polyamide	Pervaporation	Composite	–	0.27–0.95	[89]
Plasma-treated poly(4-methyl-1-pentene)coated with polyacrylic acid	Pervaporation	Composite	–	65.2	[90]
PS	UF	10 000 Da	–	1.6	[40]
Nafion 117 (perfluorinated ionomer)	Ion exchange	–	11	3.1–9.9	[76]
PVDF	MD	–	76–174	5.3–10.2	[32]
PSU/SPEEK	UF/NF	4000 Da	1.16–1.52	1.4–2.5	[91]
PES/PVP	UF	7.43–7.47 (49.83–65.19 kDa)		0.65–1.97	[83]
Polyamide hydrazide	RO	–	400–800 Å	–	[2]

[a] In nm, unless otherwise defined

Table 4.15. continued

Material	Application	Pore size [a] or MWCO (Da)	Nodule size [a]	Roughness (nm)	Reference
Polyamide of MPD/TMC	RO	–	–	0.042 – 0.84	[92]
PES/PVP	UF	9000 – 88 000 Da	–	0.38 – 1.97	[83]
Cross-linked aromatic PA composite	RO	–	–	16.9 – 82.1	[93]
Osmanic HL; Hydranautics LFC-1; Trisep X-20; Dow-FilmTec NF-70, (Thin film composite)	RO/NF	–	–	10.1 – 52.0	[94]
PEI (hf)	UF	16 – 20	91.6 – 100	2.4 – 4.4	[62]
PVDF (hf)	UF	15.63 – 48.4	85 – 174	10 – 48.4	[95]
PEI	UF	24.1 – 11.5	41.2 – 88.9	37.9 – 66.0	[31]
PEI-SMM	UF	18.1 – 9.5	47.3 – 96.5	49 – 28	[31]
PES (hf)	UF	–	54.7 – 75.2	1.21 – 2.54	[58]
Acronitrile copolymers	UF	< 500 kDa	–	3 – 28	[96]
PEI (hf)	UF	22 – 35	63	1.54 – 2.72	[63]
Sulfonated poly(ether ether ketone)	UF/NF	1.35 – 5.19 (radius)	–	0.84 – 0.93	[91]
PES/PVP	UF	40 000 and 360 000 Da	1.25 – 3.8	1.77 – 2.27	[26]
Desal G-series thin-film (Osmonics)	UF	1000 – 10 000 Da	–	3.6 – 11.7 *rms* [b]	[97]
Osmonic, DS-J	MF	3 μm	–	67	[98]
Elements G-20	UF	3500 Da	–		
Polycarbonate	UF	20 – 26	–	3.785	[99]
PES (hf)	UF	–	54.7 – 75.2	1.21 – 2.54	[58]
PSf asymmetric	UF	10 000 Da	–	1.6	[40]
PES imprinted composite membrane	MF	0.291 – 0.072 μm	–	65.72 – 96.43 *rms* [b]	[100]
PES	MF	0.1 – 10 μm	–	65.72 – 96.43 *rms* [b]	[100]
Erythrocytes	Hemodialysis	–	–	4.5 – 10.1	[101]
PS/PVP	Hemodialysis	4.5 – 7.9	38 – 16	11 – 14 *rms* [b]	[102]

[a] In nm, unless otherwise defined

[b] Root mean square of *Z* values

Table 4.15. continued

Material	Application	Pore size [a] or MWCO (Da)	Nodule size [a]	Roughness (nm)	Reference
PES/PVP	Hemodialysis	3.1 – 16.3	–	2.2 – 12.6	[66]
Glucose aldehyde cross-linked alkylatedchitosan	Biomedical Field	–	–	Below 0.1 μm	[103]
Poly(acronitrile–co-vinylchloride)	Encapsulating living cells for transplantation	–	29 – 66	–	[85]
Sulfonated PPO /PES (composite)	RO	–	72 – 32	–	[30]

[a] In nm, unless otherwise defined

References

1. Schultz R, Asunmaa S (1970) Recent Prog Surf Sci 3:291
2. Panar M, Hoehn HH, Herbert RR (1973) Macromolecules 6:777
3. Cadotte J (1981) US Patent 4,277,344
4. Cadotte J (1985) In: Lloyd DR (ed) Materials science of synthetic membranes. ACS Symposium Series 269. American Chemical Society, Washington, DC, p 273
5. Kesting RE (1990) J Polym Sci 41:2739
6. Kamide K, Iijima H, Matsuda S (1993) Polym J 25:1113
7. Wienk IM, Boomgaard Th ven den, Smolders CA (1994) J Appl Polym Sci 53:1011
8. Broens L, Altena FW, Smolders CA, Koenhn DM (1980) Desalination 32:33
9. Ray RJ, Krantz WB, Sani RL (1985) J Membr Sci 23:155
10. Kimmerle K, Strathmann H (1990) Desalination 79:283
11. Reuvers AJ, van der Berg JWA, Smolders CA (1987) J Membr Sci 34:45
12. Reuvers AJ, Smolders CA (1987) J Membr Sci 34:67
13. Pinnau I (1991) Ph.D. thesis, University of Texas
14. Boom RM, Wienk IM, Boomgaard Th ven den, Smolders CA (1992) J Membr Sci 73:277
15. Kawakami H, Mikawa M, Nagaoka S (1997) J Membr Sci 137:241
16. Khulbe KC, Kruczek B, Chowdhury G, Gagne S, Matsuura T, Verma SP (1996) J Membr Sci 111:57
17. Khulbe KC, Kruczek B, Chowdhury G, Gagne S, Matsuura T (1996) J Appl Polym Sci 59:1151
18. Pinnau I, Koros WJ (1991) J Appl Polym Sci 43:1491
19. Pfromm PH, Pinnau I, Koros WJ (1993) J Appl Polym Sci 48:2167
20. Yeh GSY, Geil P (1967) J Macromol Sci Part B Phys 1:235 and 251
21. Keith H (1965) Kolloid ZZ Polym 211:53
22. Fritzsche AK, Armuster BL, Fraundorf PB, Pelligrin CJ (1990) J Appl Polym Sci 39:1915
23. Albrecht TR, Dovek MM, Lang CA, Grutter P, Quate CF, Kuan SNJ, Frank CW, Pease RFW (1988) J Appl Phys 64:1178
24. Khulbe KC, Matsuura T, Lamarche G, Kim HJ (1997) J Membr Sci 135:211
25. Ariza MJ, Pradanos P, Rico R, Rodriguez-Castellon E, Benavente J (2003) Surf Interface Anal 35:360
26. Ochoa NA, Pradanos P, Palacio L, Pagliero C, Marchese J, Hernandez A (2001) J Membr Sci 187:227
27. Kwak SY, Yeom MO, Roh IJ, Kim DY (1997) J Membr Sci 132:183
28. Zhang W, He G, Gao P, Chen G (2003) Sep Purif Technol 30:27
29. Oh NW, Jegal J, Lee KH (2001) J Appl Polym Sci 80:2729
30. Hamza A, Chowdhury G, Matsuura T, Sourirajan S (1997) J Membr Sci 129:55
31. Khayet M, Feng CY, Matsuura T (2003) J Membr Sci 213:159
32. Khayet M, Khulbe KC, Matsuura T (2004) J Membr Sci 238:199
33. James PJ, Elliot JA, McMaster TJ, Newton JM, Elliot AMS, Hanna S, Miles MJ (2000) J Mater Sci 35:5111
34. Kasper K, Hermann KH, Dietz P, Hansma PK, Rintelen Th (1992) Ultramicroscopy 42–43:1101
35. Kawakami M, Yamashita Y, Iwamoto M, Kagawa S (1984) J Membr Sci 19:249
36. Inagaki N (1996) Plasma surface modification and plasma polymerization. Technomic, Lancaster
37. Lee KR, Teng MY, Lee HH, Lai JY (2000) J Membr Sci 164:1554
38. Won J, Kim MH, Kang YS, Park HC, Kim UY, Choi SC, Koh SK (2000) J Appl Polym Sci 75:1554
39. Mulder M (1991) Basic principles of membrane technology. Kluwer, Dordrecht
40. Vidaurre CEF, Achete CA, Gallo F, Garcia D, Simão R, Habert AC (2002) Mat Res 5:37
41. Shimomura T, Hirakawa M, Murase I, Sasaki M, Sano T (1984) J Appl Polym Sci 38:173
42. Vidaurre CEF, Achete CA, Simão R, Habert AC (2001) Nucl Instrum Methods Phys Res Sect B 175–177:732
43. Kim JY, Lee KH, Kim SC (1999) J Membr Sci 163:159
44. Dreux F, Marais S, Poncin-Epaillard F, Metayer M, Labbe M, Saiter JM (2003) Mater Res Innovations 7:183
45. Marais S, Métayer M, Labbé M, Vallenton JM, Alexander S, Saiter JM, Poncin-Epaillard F (1999) Surf Coat Technol 122:247
46. Gancarz I, Pozniak G, Bryjak M (1999) Eur Polym J 35:1419
47. Iijima Y, Tazawa T, Sato K, Oshima M, Hiraoka K (2000) Surf Interface Anal 29:596
48. Finot E, Roualdes S, Kirchner M, Rouessac V, Berjoan R, Durand J, Goudonnet JP, Cot L (2002) Appl Surf Sci 187:326
49. van't Hoff JA (1988) Ph.D. thesis, University of Twente
50. Fritzsche AK, Cruse CA, Kesting RE, Murphy MK (1990) J Appl Polym Sci 40:19

51. Fritzsche AK, Cruse CA, Kesting RE, Murphy MK (1990) J Appl Polym Sci 41:713
52. Fritzsche AK (1987) Proc Am Chem Soc Div Polym Mater Sci Eng 56:41
53. Weigel U, Schulz E, Makschin W, Albrecht W, Klug P, Gröbe V (1988) Acta Polym 39:174
54. Khulbe KC, Matsuura T (2000) J Membr Sci 171:273
55. Fujii Y, Iwatani H, Kigoshi S (1992) Polym J 24:737
56. Mahon HI (1966) US Patents 3,228,276 and 3,228,877
57. McKelvey SA (1997) Ph.D. thesis, University of Texas
58. Chung TS, Qin JJ, Huan A, Toh KC (2002) J Membr Sci 196:251
59. Kapantaidakis GC, Koops GH (2002) J Membr Sci 204:153
60. Kapantaidakis GC, Koops GH, Wessling M (2002) Desalination 145:353
61. Khulbe KC, Feng C, Matsuura T, Kapantaidakis GC, Wessling M, Koops GH (2003) J Membr Sci 226:63
62. Khulbe KC, Feng CY, Hamad F, Matsuura T, Khayet M (2004) J Membr Sci 245:191
63. Feng CY, Khulbe KC, Chowdhury G, Matsuura T, Sapkal VC (2001) J Membr Sci 189:193
64. Tsunoda N, Kokubo K, Sakai K (1999) ASAIO J 45:418
65. Asmanrafat M (2002) MSc thesis, University of Ottawa
66. Barzin J, Feng C, Khulbe KC, Matsuura T, Madaeni SS, Mirzadeh H (2004) J Membr Sci 273:77
67. Gholami MG, Nasseri S, Feng C, Matsuura T, Khulbe KC (2004) J Membr Sci 237:77
68. Hayama M, Kohori F, Sakai K (2002) J Membr Sci 197:243
69. Khulbe KC, Chowdhury G, Kruczek B, Vujosevic R, Matsuura T, Lamarche G (1997) J Membr Sci 126:115
70. Khulbe KC, Matsuura T, Noh SH (1998) J Membr Sci 145:243
71. Kesting RE (1997) Synthetic polymer membranes. McGraw-Hill, New York, p 153
72. Welty JR, Wicks CE, Wilson RE (1984) Fundamentals of momentum, heat, and mass transfer, 3rd edn. Wiley, New York
73. Nield DA (1964) J Fluid Mech 19:341
74. Scriven LE, Sterling CV (1964) J Fluid Mech 19:321
75. Koschmieder EL, Biggerstaff MI (1986) J Fluid Mech 167:49
76. Lehmani A, Durand-Vidal S, Turq P (1998) J Appl Polym Sci 68:503
77. Gebel G, Aldebert P, Pineri M (1993) Polymer 34:33
78. Soresi B, Quartarone E, Mustarelli P, Magistris A, Chiodelli G (2004) Solid State Ionics 166:383
79. Zhang Y, Shao H, Hu X (2002) J Appl Polym Sci 86:3389
80. Oh NW, Jegal J, Lee KH (2001) J Appl Polym Sci 80:2729
81. Oh NW, Jegal J, Lee KH, (2001) J Appl Polym Sci 80:1854
82. Stamatialis DF, Dias M, Pinho N (1999) J Membr Sci 160:235
83. Alsari AM, Khulbe KC, Matsuura T (2001) J Membr Sci 188:279
84. Xu X, Coleman M (1997) J Appl Polym Sci 66:459
85. Broadhead KW, Tresco PA (1998) J Membr Sci 147:235
86. Behbahani HF, Inoue H (1989) J Membr Sci 47:131
87. Reid BD, Ebron VHM, Musselman IH, Ferraris JP, Balkus KJ Jr (2002) J Membr Sci 195:181
88. Hamad F, Khulbe KC, Matsuura T (2005) J Membr Sci 256:29
89. Teng MY, Lee KR, Liaw DJ, Lin YS, Lai JY (2000) Eur Polym J 36:663
90. Wang YC, Li CL, Chang PF, Fan SC, Lee KR, Lai JY (2002) J Membr Sci 208:3
91. Bowen WR, Doneva TA, Yin HB (2002) J Membr Sci 206:417
92. Kwak SY, Ihm DW (1990) J Membr Sci 158:143
93. Hirose M, Ito H, Kameyama Y (1996) J Membr Sci 121:209
94. Vrijenhoek EM, Hong S, Elimelech M (2001) J Membr Sci 188:115
95. Khayet M, Feng CY, Khulbe KC, Matsuura T (2002) Polymer 43:3879
96. Espinoza-Gomëz H, Lin SW (2001) Polym Bull 47:297
97. Bowen WR, Doneva TA (2000) Surf Interface Anal 29:544
98. James BJ, Jing Y, Chen XD (2003) J Food Eng 60:431
99. Prádanos P, Rodriguez ML, Calvo JI, Hernández A, Tejerina F, de Saja JA (1996) J Membr Sci 117:291
100. Hilal N, Kochkodan V, Al-Khatib L, Busca G (2002) Surf Interface Anal 33:672
101. Ohta Y, Otsuka C, Okamoto H (2003) J Artif Organs 6:101
102. Hayama M, Yamamoto K, Kohori F, Sakai K (2004) J Membr Sci 234:41
103. Liu WG, Li F, Zhao XD, Yao KD, Liu QG (2002) Polym Int 51:1459

5 Pore Size, Pore Size Distribution, and Roughness at the Membrane Surface

5.1 Introduction

For many years, polymeric membranes have been widely utilized in practical applications without having precise information on their pore size and pore size distribution, despite the fact that most commercial membranes are prepared by the phase inversion technique, and the performance of those membranes is known to be governed by their pore characteristics in a complicated manner [1]. These pore characteristics are influenced both by the molecular characteristics of the polymer and by the preparative method [2]. Crudely, membranes applied for pressure-driven separation processes can be distinguished on the basis of pore diameter: as reverse osmosis (RO, < 1 nm), dialysis (2–5 nm), ultrafiltration (UF, 2–100 nm), and microfiltration (MF, 100 nm to 2 μm). Nanofiltration (NF) membranes are a relatively new class and have applications in a wide range of fields [3]. The pore sizes of NF lie between those of RO and UF membranes.

The characteristics of membrane pore structures (pore size, pore size distribution, pore density, surface roughness, etc.) should be the backbone of the membrane industry, since such characteristics govern the filtration properties of membranes. Hence, Smolders and Vugteveen [4] and Zeman and Tkacik [5] discussed a number of methods for determining the physical characteristics of skinned UF membranes, including their pore size and pore size distributions. There are a number of ways to measure the pore size and the pore size distribution [6, 7], e.g., the bubble point method, mercury porosimetry, thermoporometry, permporometry, and the adsorption method, as well as methods based on liquid or gas transport, microscopic methods such as scanning electron microscopy, transmission electron microscopy, and AFM. AFM is a novel technique, and its application to membranes, both biological and synthetic, is growing rapidly [8, 9].

Interestingly, the pore sizes obtained from SEM and TEM are generally smaller than those obtained from AFM images. There may be a number of reasons for this as suggested by Bowen et al. [10]. SEM requires deposition of a conducting coat on the sample, and TEM requires preparation of a replica. Structural changes may also occur due to damage by the electron beam or the requirement to operate in a high vacuum [11]. AFM has a tremendous advantage because of its operation in air with no sample preparation. The only operational requirement is attachment of a membrane sample to a steel disc with double-sided tape. In SEM, we obtain a *visual represen-*

tation of the membrane structure. Analysis of the photomicrographs yields the pore size distribution. On the other hand, some numerical parameters can be obtained directly from AFM pictures [12], e.g. the pore size and pore size distribution, surface pore density (the number of pores per unit area) and porosity (the porous surface fraction).

5.1.1 Porous Structure of the Membrane Surface, SEM

There are a number of methods to characterize the membrane pore structure. Among those, SEM is one of the most popular. SEM provides two-dimensional images of surfaces. One such example is given in Fig. 5.1 for the surfaces of polysulfone membranes [11]. Figure 5.1a shows the membrane when the cast film was coagulated in a water bath, while Fig. 5.1b shows the surface when the cast film was coagulated in a 20/80 mixture of water/*N*-methyl-2-pyrrolidone (NMP). In Fig. 5.1a, very small domains exist that cannot be clearly defined as either nodules or interconnected cavity channels due to the low resolution of SEM. On the other hand, Fig. 5.1b shows that pores of various sizes exist on the membrane surface. The average pore diameter is 112 nm. Similar SEM images are reported in the literature for flat sheet membranes [13–15].

SEM usually underestimates pore diameters due to the metal coating that is necessary to increase conductivity. Pore diameter is varied with coating rate, coating period, and pore shape. The pore shape is usually not cylindrical but funnel-shaped, so coating can reduce the pore size. Structural changes may also occur due to the damage by the electron beam or by the requirement to operate in a high vacuum. SEM shows the defects with minimal information on the surrounding surface depression due to the two-dimensional character of the image (i.e., the interconnected cavity channels).

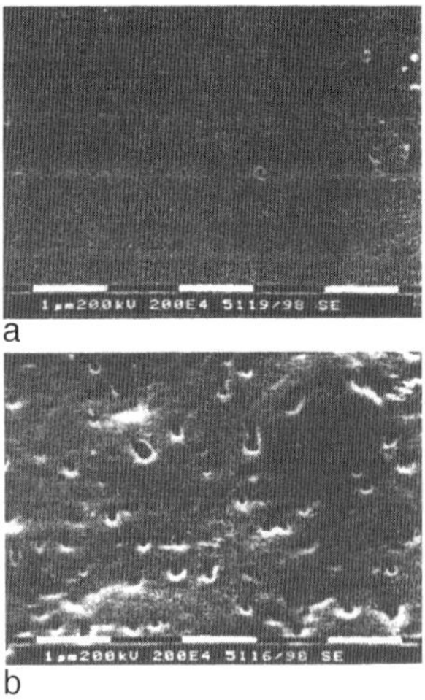

Fig. 5.1a,b. SEM photograph of PSf membrane surface. **a** Sample 1. **b** Sample 2. Reprinted from [11]. Copyright 1999, with kind permission from Elsevier

5.1.2 Porous Structure of Membrane Surface, AFM

The atomic force microscopy technique is now widely used for the study of membrane surfaces. It has become an important tool of imaging the surface of materials to atomic-level resolution, and this technique does not need any special sample preparation, which is essential for SEM and TEM. AFM can show three-dimensional images of the surfaces. Paredes et al. has written an excellent review on the application of AFM for the characterization of microporous and mesoporous materials [16].

It is interesting to compare the images of the same sample taken by both SEM and AFM. The surface structures of the PSf membranes shown in Fig. 5.1 are also imaged by AFM and shown in Fig. 5.2 [11]. Unlike Fig. 5.1a, Fig. 5.2a shows a typical nodular structure and interconnected cavity channels between the agglomerated nodules. A totally different surface structure exists in the case of sample 2 (Fig. 5.2b). There is no nodular structure, but large pores exist throughout the whole membrane surface. This example demonstrates that the AFM images can show the membrane surface in much more detail than SEM images.

The AFM can also give high-resolution images of a surface in air and even under liquids [13,14,17]. Dietz et al. studied PSf UF membranes in air and under water [17]. Figure 5.3 shows the AFM images of a PSf membrane with a molecular weight cutoff (MWCO) of 10 000 Da in air (Fig. 5.3a and b) and under water (Fig. 5.3c). When the membrane is dry, the pore density is 350 pores μm^{-2} and the average distance between the single pores is 60 nm (Fig. 5.3a). Single pores with diameters between 15 and 30 nm are clearly visible. Smaller pores could not be measured because the tips were larger than the pores and the tips were prevented from moving deep enough into the pores. In the early stage, when water was added to the surface of the sample, it was not possible to take the AFM image due to the starting of the swelling process. After one hour, when the swelling process ceased, the reproducible images could be

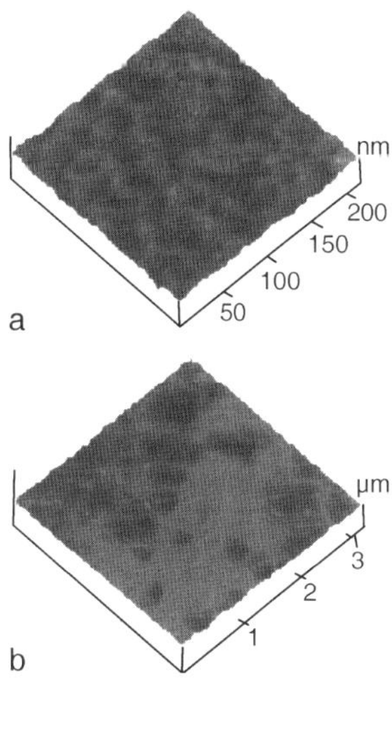

Fig. 5.2a,b. Three-dimensional tapping mode AFM image of PSf membrane surface. **a** Sample 1. **b** Sample 2. Reprinted from [11]. Copyright 1999, with kind permission from Elsevier

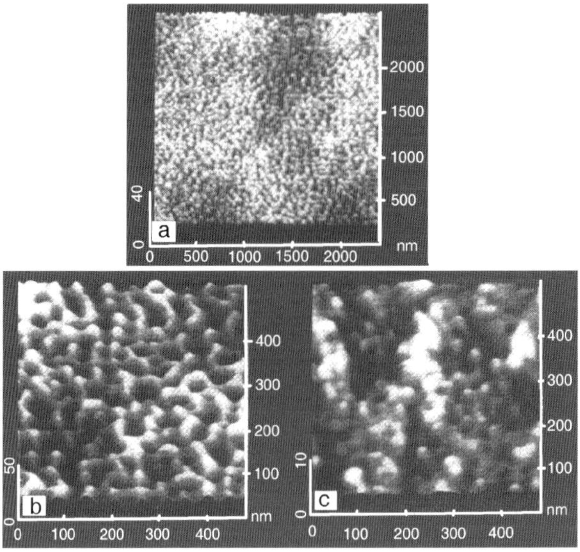

Fig. 5.3a–c. a Polysulfone ultrafiltration membrane with cutoff value of 10 000 (UFM2, Millipore PTGC) in air. **b** Zoomed-in scan from area in **a**. **c** The same membrane surface under water. Reprinted from [17]. Copyright 1991, with kind permission from Elsevier

obtained. The surface structure changed from that of the surface in the air (compare Fig. 5.3b and c); in water, the surface is much more corrugated, and the pores are smaller.

5.2 Pore Size and Pore Size Distribution at the Membrane Surface

5.2.1 Determination of Pore Size and Pore Size Distribution by AFM

The pore size and the pore size distribution of membranes can be determined using AFM in both the contact and the tapping mode in air, and by the contact mode in liquid [18–20]. The ability to measure the size of pores by AFM can obviously be enhanced when a good image is produced. In this field, the school of Bowen, Swansea, UK, has made a remarkable achievement.

The pore size and pore size distribution according to the log-normal distribution can be determined by the method of Singh et al. [20] using AFM images. The sizes of the pores that most likely represent the opening of the pores can be measured by visual inspection of the line profile (Fig. 3.5). Note that there are several pores involved in one such profile obtained at different areas of a membrane. For elongated pores, the largest dimensions were assigned as the effective pore diameters.

The pore sizes so measured are arranged in an ascending order. Median ranks are calculated from the following equation:

$$\text{\textit{Median or} } 50\% \text{ \textit{rank}} = [(j - 0.3)/(n + 0.4)] \times 100 \tag{5.1}$$

where j is the order number of the pore when arranged in ascending order and n is the total number of pores measured.

To obtain a cumulative distribution graph, these median ranks are plotted on the ordinate against pore sizes arranged in an increasing order on the abscissa. This plot will yield a straight line on a log-normal probability paper, if pore sizes have a log-normal distribution, as shown in Fig. 5.4 [20]. From this graph, values of mean pore sizes, μ_p, and geometric standard deviation, σ_p, can be calculated.

For the measurement of pore sizes on the membrane, the geometry of the tip is also important. The pore sizes or funnel shape of pore entrances determined by AFM depend on the convolution between tip shape and pore shape as illustrated in Fig. 5.5. In this figure, AFM traces for two different membrane models (cylindrical and funnel-shaped with two different pore sizes) and an idealized round-shaped tip. Only in model A can surface diameter be measured accurately from the tip traces.

Bowen et al. [10,19,21] investigated the surface pore structure of Cyclopore and Anopore in air. They used both contact and noncontact mode and showed the same features in both modes. Pore dimensions obtained by both techniques were also in

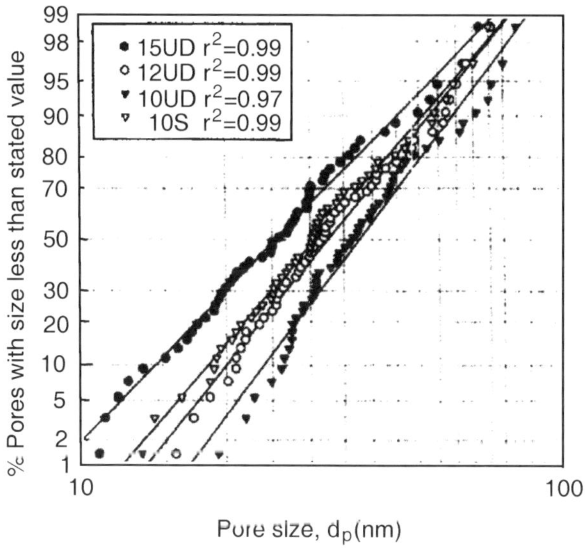

Fig. 5.4. Log-normal pore size distribution measured from AFM images. Reprinted from [20]. Copyright 1998, with kind permission from Elsevier

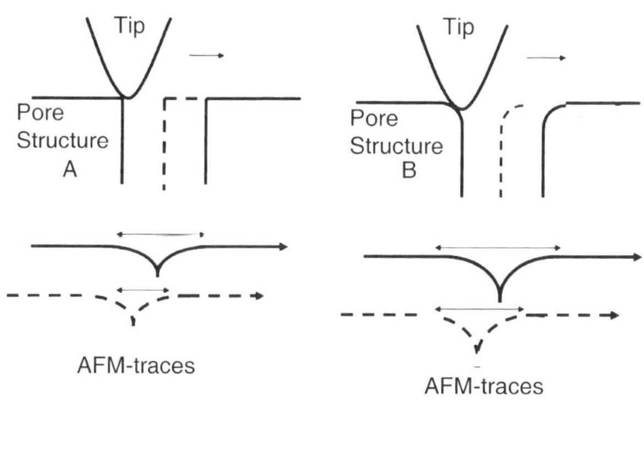

Fig. 5.5. Interaction between tip and pore structure. AFM images of membrane pores have to be interpreted as a convolution between tip shape and pore shape (if the sizes are comparable)

very good agreement. For example, the mean pore diameter obtained for a Cyclopore membrane (nominal pore size 0.2 μm) imaged in contact mode was 0.196 μm with standard deviation of 0.047 μm compared to 0.186 μm with a standard deviation of 0.033 μm by noncontact mode.

Bowen et al. have also carried out a very comprehensive AFM investigation on porosity covering many different types of polymeric membranes [10]. Membranes made from poly(ether sulfone), regenerated cellulose, polysulfone, and poly(acrylonitrile-vinylchloride), among others, were investigated by these researchers [10,18,19, 22,23]. Their work concluded that it was possible to obtain quantitative information on pore structures and pore size distributions by analyzing AFM images obtained in the tapping mode of MF and UF membranes. These studies indicated that TM-AFM is better than contact mode AFM as it provides better pore definition in membranes with small pores [18] and damage can be avoided even though the polymers are soft materials. Hilal et al. [24] wrote a brief review on using AFM toward the improvement in NF membranes. In this article, the authors presented a brief review on the potential use of the *state-of-the-art* AFM technique, as a method for surface characterization, to understand membrane characteristics so as to significantly improve NF membrane properties. They also predicted that the AFM technique would allow the effect of surface roughness on transmembrane transport and the fouling potential of NF membranes to be quantified. Figure 5.6 shows AFM images of membrane surfaces with pores of nanometer dimensions [24].

Bowen et al. [18] investigated the surface pore structure of a poly(ether sulfone) UF membrane with a MWCO of 25 000 (ES625 manufactured by PCI Membrane Systems). Figure 5.7 shows an AFM image of the membrane in three-dimensional form over an area of 100 × 100 nm with the light regions being the highest points and the darkest regions being the pores. Figure 5.8 shows a high-resolution, three-dimensional image over an area of 14 × 14 nm. (The scan size was 25 × 25 nm, and the area containing the pore has been selected.) A clear image of a single pore in the membrane is visible. Further analysis of an image involving 111 pores gave an average pore size of 5.1 nm with a standard deviation of 1.1 nm. The range of pore sizes was 3.1–7.7 nm.

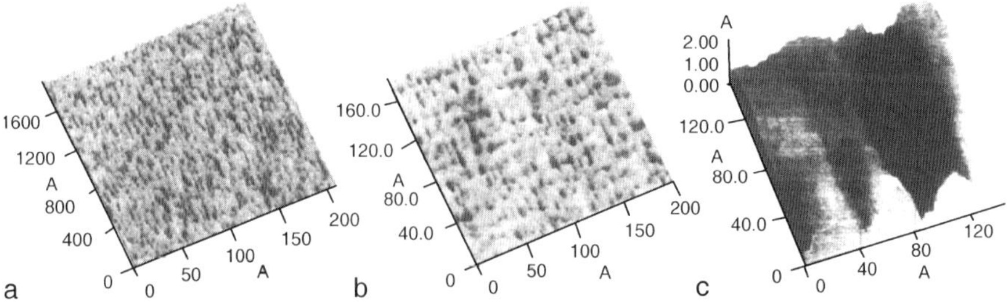

Fig. 5.6a–c. Visualization of membrane surfaces. **a** AFM image of an ES404 membrane (MWCO 4000). **b** AFM image of a modified XP117 membrane (MWCO 40 000). **c** AFM image of a single pore of 4 nm in an NF membrane. Reprinted from [24]. Copyright 2003, with kind permission from Elsevier

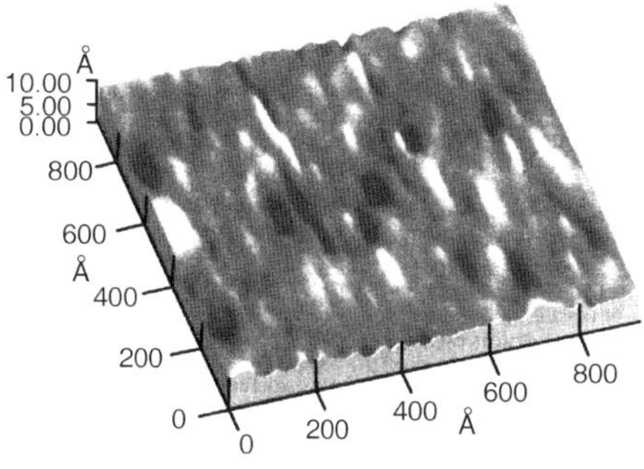

Fig. 5.7. Three-dimesional non-contact AFM image of an ES625 ultrafiltration membrane. Reprinted from [18]. Copyright 1996, with kind permission from Elsevier

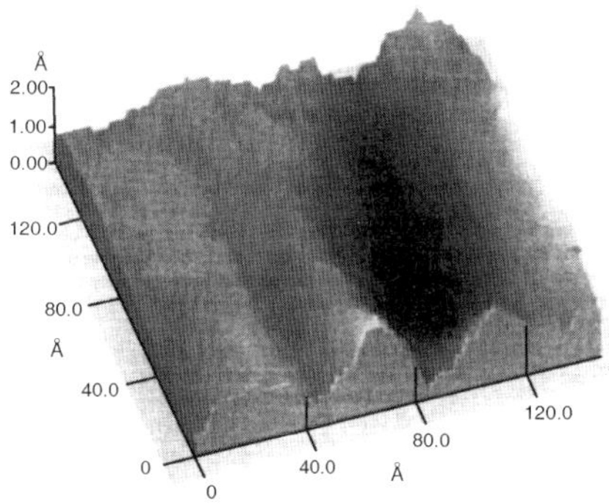

Fig. 5.8. Three-dimensional non-contact AFM image of a single pore in an ES625 ultrafiltration membrane. Reprinted from [18]. Copyright 1996, with kind permission from Elsevier

In another study on three Cyclopore membranes C01, C02, and C04 (Whatman International Limited) with a nominal pore size of 0.1, 0.2, and 0.4 μm, respectively, Bowen et al. clearly observed pores [10, 22]. Figure 5.9a–c shows the AFM images over an area of 3 × 3 μm. The AFM software allowed quantitative determination of the diameter of the pores by use of the images in conjunction with digitally stored line profiles. Comparing these membranes at higher magnification showed the approximately circular shape of the pore entrances and the increase in pore size from C01 to C04. Bowen et al. [10] also studied four YM Diaflo membranes (Amicon Inc., USA) as examples for UF membranes. They are YM3, YM10, YM30, and YM100 membranes with MWCO of 3000; 10 000; 30 000; and 100 000 Da, respectively. Figure 5.10a–d shows three-dimensional AFM images of these membranes. The pores are clearly visible as small, well-defined dark areas of the image. In some cases they appear to occur in clusters, and for the higher MWCO membranes, there is an increasing tendency for the pores to occur at the crest in the membrane surface. Table 5.1 shows statistical information on the mean pore diameter (size), the standard deviation, and the size range. The mean pore diameter increases systematically as the specified MWCO of the membrane increases, with a small standard deviation in all cases [10].

Bowen's school has successfully used the AFM images of UF membranes and produced accurate surface statistics. In some cases, UF membranes may also have pores of subnanometer dimensions. Figure 5.11 shows a high-resolution image of a single pore of around 0.5 nm in an XP117 membrane (MWCO 4000, PCI Membrane Systems Ltd.) [10].

Table 5.1. Statistical characterization of Diaflo membranes

Diaflo membrane	MWCO	Mean pore diameter (nm)	Standard deviation (nm)	Size range (nm)	Number of counted pores	R_q [a] (nm)
YM3	3000	8.7	1.3	6.2 – 12.3	96	0.36
YM10	10 000	11.3	2.4	6.1 – 18.0	90	0.53
YM30	30 000	13.2	2.8	8.2 – 22.6	80	0.6
YM100	100 000	19.4	4.2	10.1 – 32.4	78	0.77

[a] R_q root mean square

Table 5.2. Data from membranes used by Dietz et al. [9]

Membranes	Manufacturers	Polymer	MWCO /pore size
PCTE 10	Poretics [1]	Polycarbonate	10 nm
PCTE 50	Poretics	Polycarbonate	50 nm
PCTE 100	Poretics	Polycarbonate	100 nm
PTHK	Millipore [2]	Polysulfone	100 000 Da
PTGC	Millipore	Polysulfone	10 000 Da
DUS-1020	Celfa [3]	Poly(ether sulfone)	100 000 Da
DUS-0420	Celfa	Poly(ether sulfone)	40 000 Da
DUS-K520	Celfa	Poly(ether sulfone)	5000 Da
SM14669	Sartorius [4]	Polysulfone	100 000 Da
SM14639	Sartorius	Polysulfone	10 000 Da
SM14639s [5]	Sartorius	Polysulfone	10 000 Da

[1] Poretics Corp., Livermore, CA (USA)
[2] Millipore Inc., Bedford, MA (USA)
[3] Celfa AG, Seewen (Switzerland)
[4] Sartorius AG, Göttingen (Germany)
[5] Especially smooth surface (manufacturer specification)

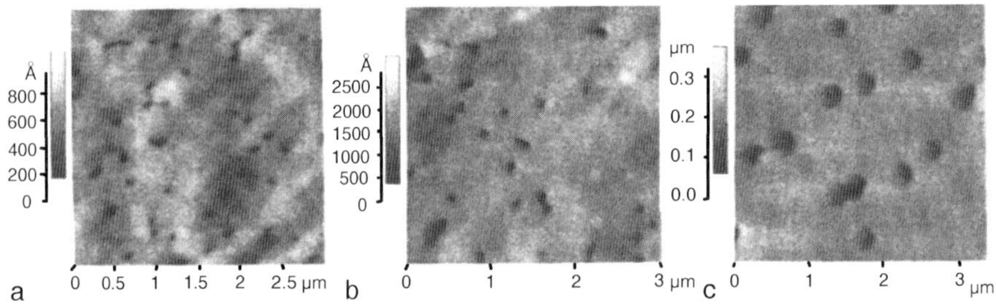

Fig. 5.9a–c. Two-dimensional images of **a** C01, **b** C02, and **c** C04 Cyclopore membranes. Reprinted from [10]. Copyright 1999, with kind permission from Dekker

Dietz et al. [9] studied AFM images of three capillary pore membranes and eight UF membranes. The membranes imaged by Dietz et al. are listed in Table 5.2.

Figure 5.12 shows AFM images of the surface of three track-etched capillary pore membranes with nominal pore diameters of 10, 50, and 100 nm. Differences in the pore sizes of these three membranes are clearly visible in the AFM pictures. However,

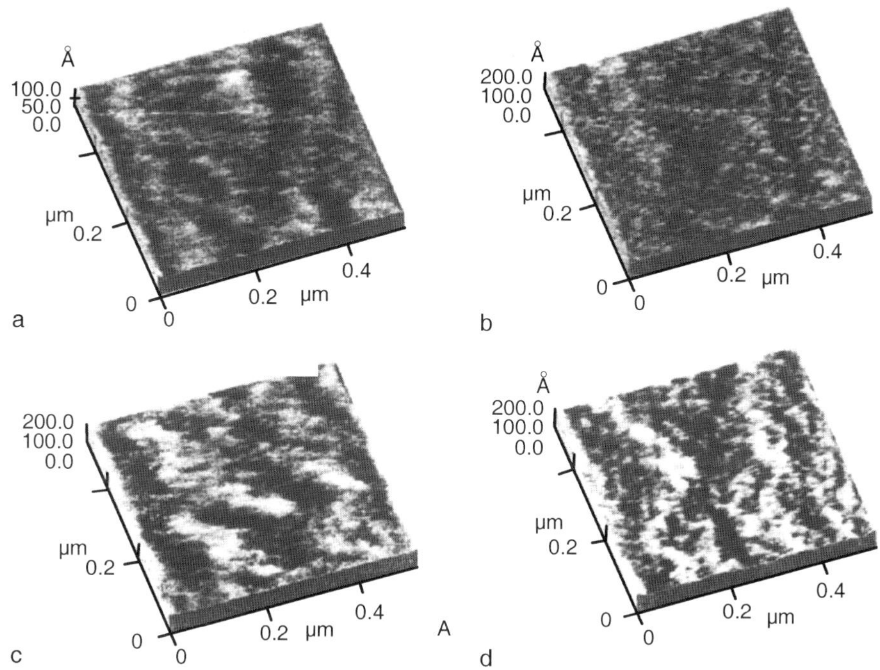

Fig. 5.10a–d. Three-dimensional images of YM Diaflo membranes: **a** YM3, **b** YM10, **c** YM30, and **d** YM100. Reprinted from [10]. Copyright 1999, with kind permission from Dekker

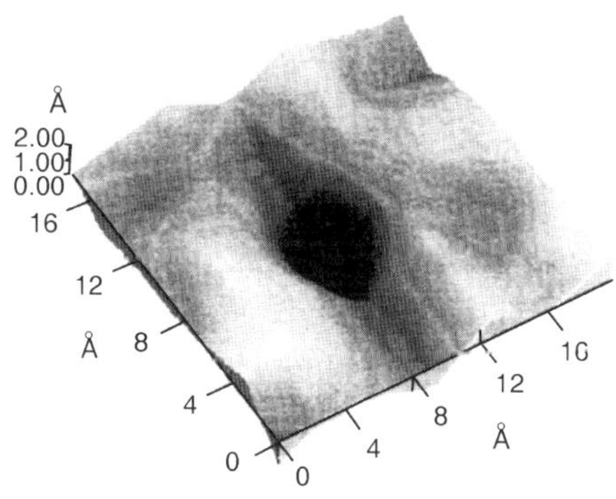

Fig. 5.11. Three-dimensional image of a 0.5 nm single pore in an XP117 membrane (MWCO 4000). Reprinted from [10]. Copyright 1999, with kind permission from Dekker

10 nm pores are difficult to recognize because the surface of this membrane is highly corrugated. It is also evident that the pore diameters vary within each membrane, and the pore distribution is strongly inhomogeneous due to the statistical process of fabrication [9]. At the bottom of each picture, an 800-nm-long line profile across one typical opening is presented.

Figure 5.13a–h shows AFM images of the eight UF membranes. All images are given in topographical representation with a view angle of 60°, which emphasizes the three-dimensional character of the AFM images. Higher magnification images

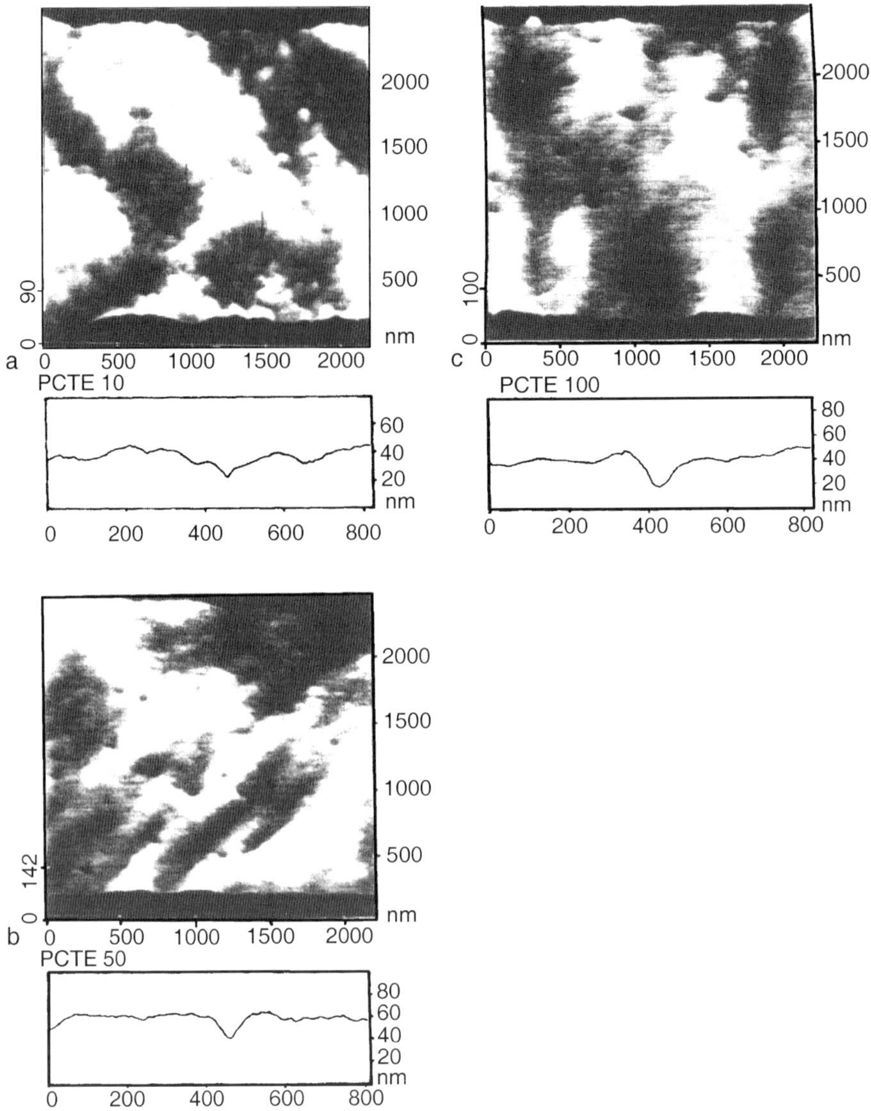

Fig. 5.12a–c. AFM images of three track-etched capillary pore membranes with nominal pore sizes: **a** 10 nm, **b** 50 nm, and **c** 100 nm. Two of these pores in PCTE 10 are indicated by *arrows*. The bottom parts show 800-nm-long line profiles across typical pore opening. Reprinted from [9]. Copyright 1992, with kind permission from Elsevier

of PTHK, PTGC, DUS-1020, DUS-K520, SM14669, and SM14639, together with line profiles, are shown in Fig. 5.14a–f, which enables the study and comparison of single pores. The surface structure shows significant differences from membrane to membrane. No surface can be considered smooth on the molecular scale. PTHK and PTGC show a fairly uniform and rather similar appearance with agglomerated nod-

Table 5.3. Pore characteristics of MF and UF membranes

Membranes	Mean surface pore diameter (nm)	Pore density (pores μm^{-2})	Surface porosity (%)
PCTE 10	18.1	5.2	0.6
PCTE 50	65.7	5.2	1.4
PCTE 100	113.3	4.8	4.5
PTHK	22.1	88	3.4
PTGC	14.1	435	6.8
DUS-1020	25.2	128	6.3
DUS-0420	11.6	482	5.1
SM14669	< 26.2*	120	< 6.5*
SM14639	12.6	172	2.2
SM14639s	18.8	144	4.0

* Upper limit

ules, but with a very different number of pores between these nodules. Values of pore sizes, pore densities, and porosities were determined by analysis of the AFM images as summarized in Table 5.3. No data are given for DUS-K520 due to its indistinct pore structure; the large openings cannot be considered single pores [9]. For SM 14669, the reported values are only upper limits because some of the large openings are probably composed of two or more pores.

Surface pore diameters were measured by visual inspection of the line profiles of 50 pores of each membrane. All membranes have a wide pore size distribution. The deviation between ±20% and ±40% from the average value is noticeable in most cases and is higher for membranes with larger pores (or higher MWCOs). The *pore density* was obtained by observing several AFM images from different sample areas of the same membrane and counting the number of pores in a unit area. Surface porosity is defined as the ratio of the pore area to the total area of the membrane. The porosity is low and varies between 0.6% and 7%. No relationship between MWCO and porosity was found.

Hilal et al. [25] studied the surface structure of molecularly imprinted poly(ether sulfone) membranes (called MIP membranes) by AFM and quantified the pore size and the surface roughness. They modified PES microfiltration membranes with a normal pore diameter of 0.22 μm and a thickness of 150 μm (Millipore). First, the membranes were coated with photoinitiator by soaking them in a 0.25 M solution of benzoin ethyl ether (BEE) in methanol and then immersing them in a mixture of 80 mM trimethyl propane trimethacrylate (TRIM), 40 mM 2-hydroxyethyl methacrylate (HEMA), and 2 mM adenosine 3',5'-cyclic monophosphate (cAMP) in an ethanol–water mixture (70:30 vol.%). Thereafter, the membranes were exposed to a B-100 lamp of relative radiation intensity 21.7 mW cm^{-2} at 355 nm. Membranes with different modifications were obtained using various UV exposure times. The residual nongrafted polymer, monomer, initiator, and the template were extracted with methanol. After drying, the degree of modification (*DM*) was calculated from the

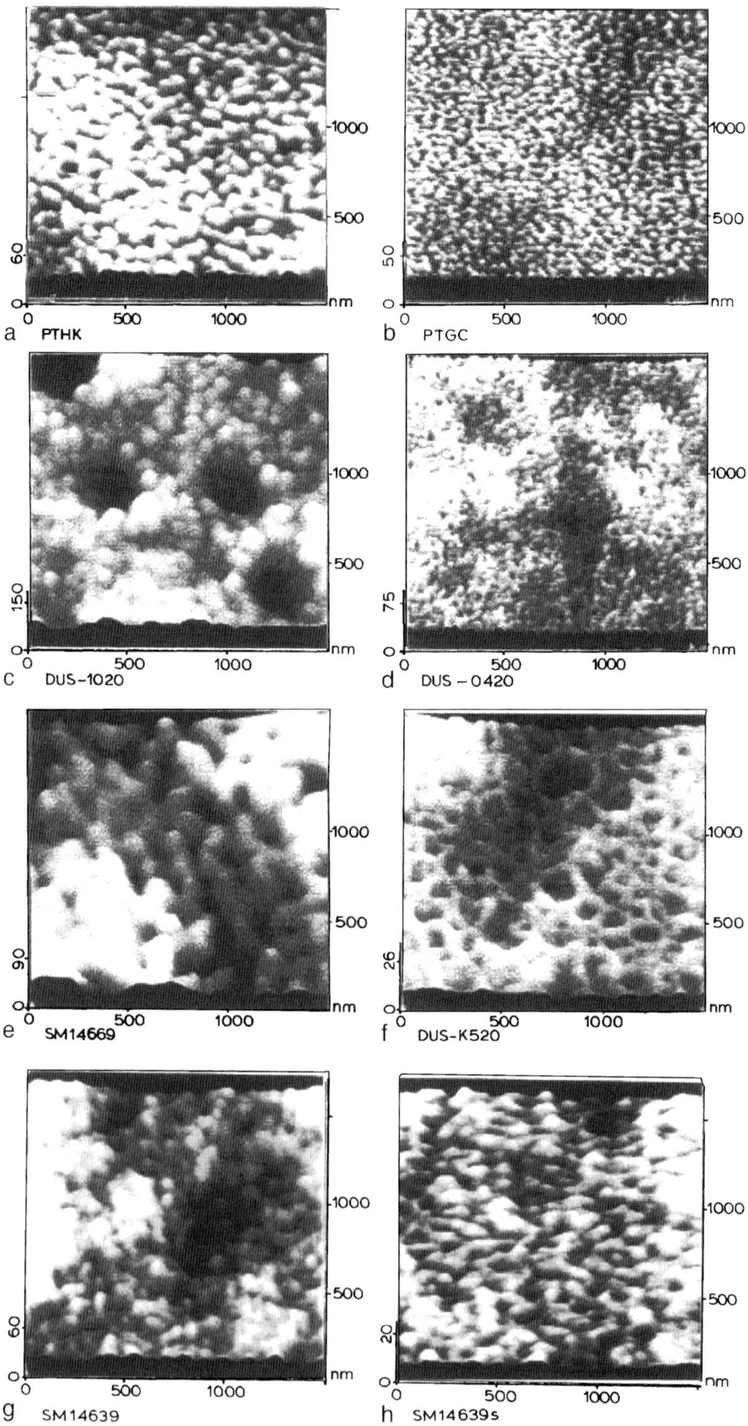
a PTHK
b PTGC
c DUS-1020
d DUS – 0420
e SM14669
f DUS-K520
g SM14639
h SM14639s
nm
0
500
1000

Fig. 5.13a–h. AFM images of different UF membrane surfaces: **a** PTHK, **b** PTGC, **c** DUS-1020, **d** DUS-0420, **e** SM14669, **f** DUS-K520, **g** SM14639, and **h** SM14639s. Note: The scan area is the same for all images (1.5 × 1.5 μm^2), but the *z*-scale differs considerably and is shown on the left of the images. Reprinted from [9]. Copyright 1992, with kind permission from Elsevier

Table 5.4. AFM characteristics of the porous structure of MIP membranes with various degrees of modification

Membranes (degree of modification)	Mean pore diameter by AFM (μm)	$R_{p\text{-}v}$* (nm)	R_q** (nm)
PES (conventional)	0.288	159.51	65.72
PES (260 μg cm^{-2})	0.247	123.46	72.77
PES (460 μg cm^{-2})	0.291	139.11	83.31
PES (640 μg cm^{-2})	0.072	223.54	96.43

* Peak-to-valley roughness
** Root mean square

Table 5.5. Pore diameters for UF and MF membranes obtained from filtration experiments and AFM images

Membranes	Pore size, diameter (nm)		
	Filtration	AFM	
		Small	Large
40 kDa	11.0 ± 1.0	–	–
100 kDa	15.4 ± 1.0	31 ± 10	69 ± 20
200 kDa	28.4 ± 1.0	38 ± 10	114 ± 20
0.1 μm	80.0 ± 1.0	96 ± 10	185 ± 60

weight difference of the modified and conventional membrane. Blank membranes were prepared using the same procedure, but without using the template for comparison.

Atomic force microscopy images clearly indicated that a consistent increase in the degree of modification led to a systematic decrease in pore size and an increase in surface roughness (Table 5.4). The AFM characteristics of imprinted membranes are in good correlation with the filtration data. Figure 5.15a shows a high-resolution AFM image of a conventional membrane (without modification) in three-dimensional form over an area 3.5 × 3.5 μm. The pores are clearly visible as small, well-defined dark areas. Figure 5.15b–d shows three-dimensional images with different degrees of modification: 260, 460, and 640 μg cm^{-2}, respectively.

Bessières et al. [26] studied the surfaces of sulfonated polysulfone (SPS) membranes with MWCOs of 40, 100, and 200 kDa and a PVDF membrane with a pore size of 0.1 μm by AFM. Table 5.5 shows the small- and large-pore diameters for the above UF and MF membranes obtained from AFM images together with the pore sizes obtained from filtration experiments.

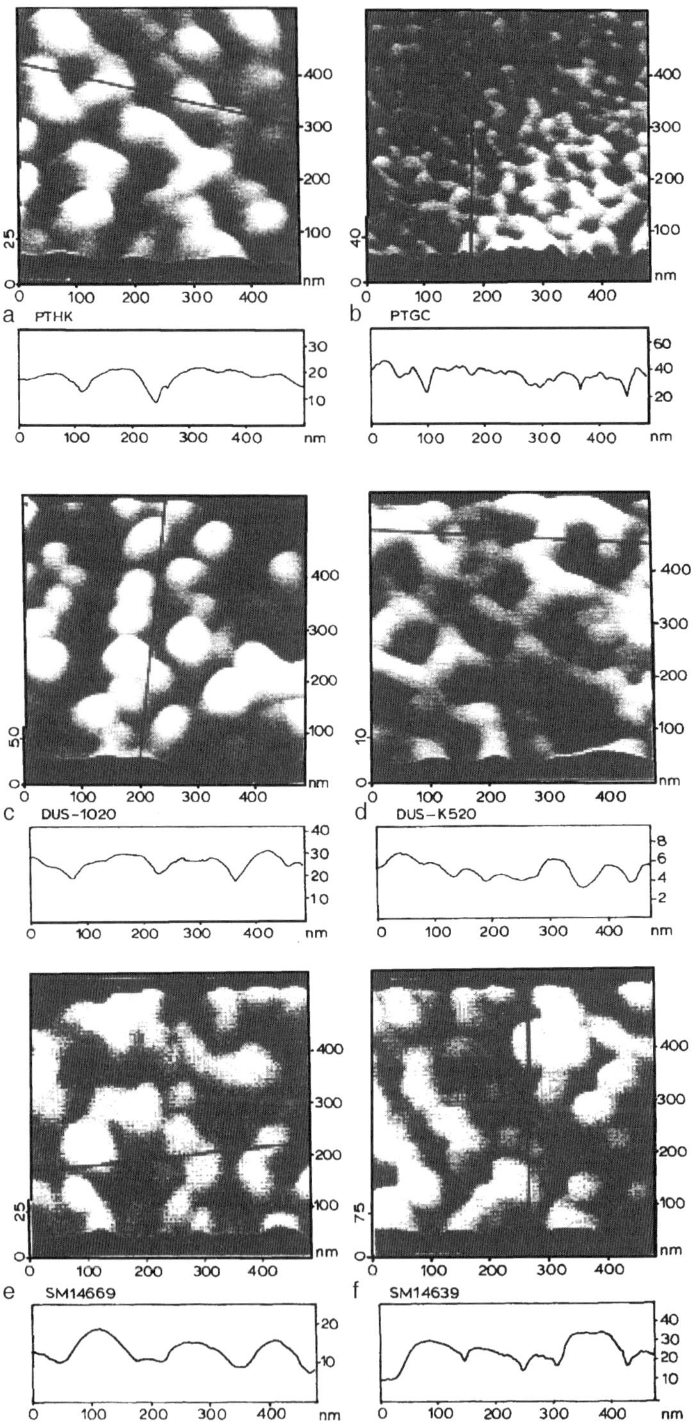
a PTHK
b PTGC
c DUS-1020
d DUS-K520
e SM14669
f SM14639

Fig. 5.14. Higher magnifications (470 × 470 nm) of some of the UF membranes (PTHK, PTGC, DUS-1020, DUS-K520, SM14669, and SM14639). The cross sections in the bottom part of each picture were taken along the indicated lines and include profiles of typical pore openings. These profiles were used to determine pore diameter. Reprinted from [9]. Copyright 1992, with kind permission from Elsevier

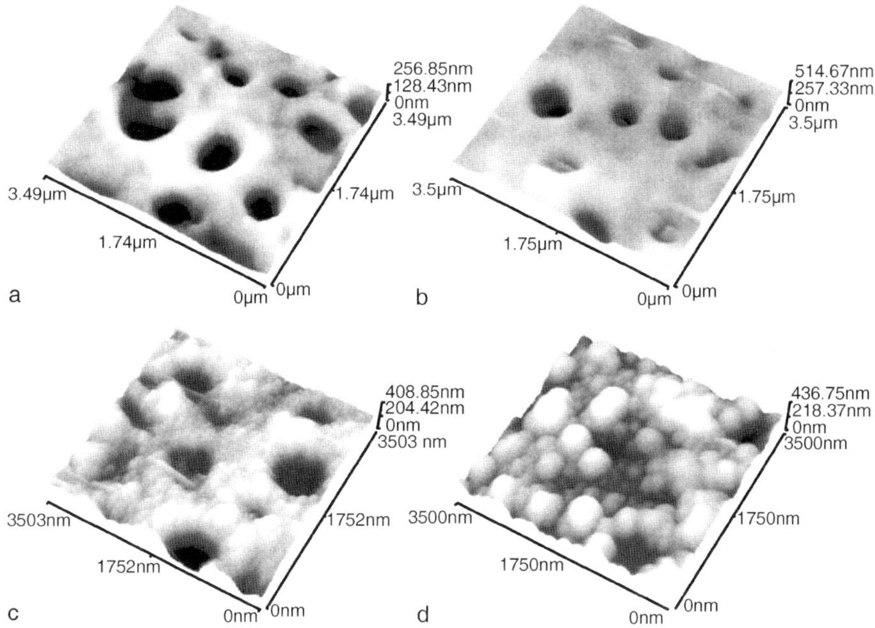

Fig. 5.15a–d. Three-dimensional AFM images of PES membranes. **a** Conventional. **b** Imprinted with degree of modification (*DM*) 260 μg cm^{-2}. **c** Imprinted with *DM* 460 μg cm^{-2}. **d** Imprinted with *DM* 640 μg cm^{-2}. Reprinted from [25]. Copyright 2002, with kind permission from Wiley

Gordano et al. [27] characterized three different composite HYFLON AD 60X membranes by AFM. Copolymers of tetrafluoroethylene (TFE) and 2,2,4-trifluoro-5-trifluoromethoxy-1,3 dioxole (TTD), as well as amorphous perfluoropolymers commercially known as HYFLON® AD, were used for the preparation of composite membranes. The support for the composite membranes was polyamide (PA) MF membranes from AKZO with a normal pore size of 0.2 μm. After coating a thin layer of copolymer solution (1% wt/wt) on the supporting membrane, the solvent was evaporated. The evaporation temperature was −5, 4, and 25 °C. After drying the membrane at the required temperatures for 24 h, the membrane was further dried at room temperature for 24 h. The AFM images of these membranes were obtained in 0.1 M NaCl solution. Fast Fourier transform filtering was applied to all images to remove unwanted noise and to improve resolution [22]. The results of the AFM image analysis are given in Table 5.6.

Table 5.6. Surface characteristics of membranes [27]

Temperature for solvent evaporation (°C)	Average pore diameter (nm)	R_q [a] (nm)
−5	18	15.2 (±3.1)
4	25	8.6 (±0.1)
25	41	3.1 (±0.4)
PA	20	18.3 (±0.9)

[a] Root mean square (numbers in parentheses are standard deviations; *rms* roughness measured over an area of 1 × 1 μm)

The surface roughness of the PA membrane (support) was reduced remarkably on coating. The roughness parameters of the top surface of the membranes cast at −5 °C are greater than those on other coated membrane surfaces.

5.2.2 Comparison with Other Methods

Singh et al. [20] calculated the pore sizes of asymmetric poly(ether sulfone) membranes used for UF experiments. The membranes were made by the phase inversion technique using casting solutions of different PES concentrations (12, 15, and 20 wt.%) in *N*-methyl-2-pyrrolidone (NMP). They measured the pore sizes by solute transport and also calculated them from AFM images. Their results are given in Table 5.7, which indicates that average pore sizes of membranes calculated by AFM images are almost three times higher than those calculated from solute transport data.

Hayama et al. [28] studied a hollow fiber dialysis membrane APS-150 (Asahi-Medical, Japan) made of polysulfone (PSf) (asymmetric structure) by using field emission scanning electron microscopy (FE-SEM) and AFM. The hollow fiber had an inner diameter of 210 μm and a wall thickness of 45 μm. Figure 5.16a shows the results of observation of sectional, inside, and outside surfaces of the dry APS hollow fiber dialysis membrane by FE-SEM. Figure 5.16b shows the pore diameter distribution on the inside and outside surfaces by image analysis. The thickness of the coated metal film was approximately 12 nm, thus, the small pores were buried and became invisible, and the larger pores were reduced in size. Figure 5.17a and b shows the AFM image of inside and outside surfaces using a normal silicon single-crystal probe (radius of curvature 5–20 nm, probe NCH). Pores on the outside surface were clearly

Table 5.7. Average pore size and geometric standard deviation for various PES UF membranes calculated from solute separation data and from AFM images

PES in casting solution (wt.%)	From solute transport Average pore size (nm)	Geometric std. dev.	From AFM images Average pore size (nm)	Geometric std. dev.
10	10.38	1.78	37.6	1.43
12	9.14	1.74	32.4	1.46
15	7.18	1.84	25.4	1.57

observed while this probe failed to detect clearly the pores on the inside surface. On using a sharpened probe (radius of curvature 2 nm, probe SSS-NCH), it was possible to observe the pores less than 5 nm in diameter. Figure 5.18 shows the distribution of the pore diameters determined by TM-AFM. Table 5.8 shows a comparison of average pore diameters obtained by different techniques.

The average pore diameter calculated by the Hagen-Poiseuille equation lies between the values for the outside and inside surfaces obtained by TM-AFM. This is because the Hagen-Poiseuille equation measures the pore diameter somewhere along the permeation pathway through which water travels [28]. On the other hand, the pore diameters measured by FE-SEM and TM-AFM are the values on the membrane surface.

Mohammad et al. [29] fabricated NF composite membranes by the interfacial polymerization technique and studied the membrane's surface by AFM. The membrane support was prepared from a dope containing polysulfone (PSf) (P1835-BP Amoco) and poly(vinylpyrrolidone) (PVP) (Fluka) with *N*-methyl-2-pyrrolidone (NMP) as the solvent. The top active layer was obtained through interfacial polymerization between trimesoyl chloride (TMC) in hexane and the aqueous phase containing bisphenol A (BPA). Table 5.9 shows the summary of the membrane preparation conditions. The first three membranes identified as PT-30, PT-45, and PT-60 differ in the period of interfacial reaction. The other three membranes identified as PC-05, PC-1, and PC-2 differ in terms of the concentration of BPA in the aqueous phase. The pore sizes determined by AFM and also calculated using the Donnan-steric-

Table 5.8. Comparison of average pore diameters of APS-150 hollow fiber membranes determined by FE-SEM, TM-AFM, and Hagen-Poiseuille equation

Observed portion	Average pore diameter (nm)			
	FE-SEM	TM-AFM with NCH [a]	TM-AFM with SSS-NCH [b]	Hagen-Poiseuille equation
Inside	14.4	17.4	15.8	24.8
Outside	566	715	–	–

[a] Probe radius curvature 5 – 20 nm
[b] Probe radius curvature 2 nm

Table 5.9. Code names of membranes, reaction times, BPA concentrations used for the particular membrane preparation, and pore sizes determined by AFM and calculated by DSPM modeling

Membranes	Reaction time (s)	BPA concentration (wt.%)	Pore diameter by AFM (nm)	Pore diameter by DSPM (nm)
PT-30	30	1	1.68	1.39
PT-45	45	1	1.57	1.47
PT-60	60	1	1.21	1.45
PC-05	45	0.5	5.36	1.38
PC-1	45	1.0	1.57	1.47
PC-2	45	–	< 0.4	2.36

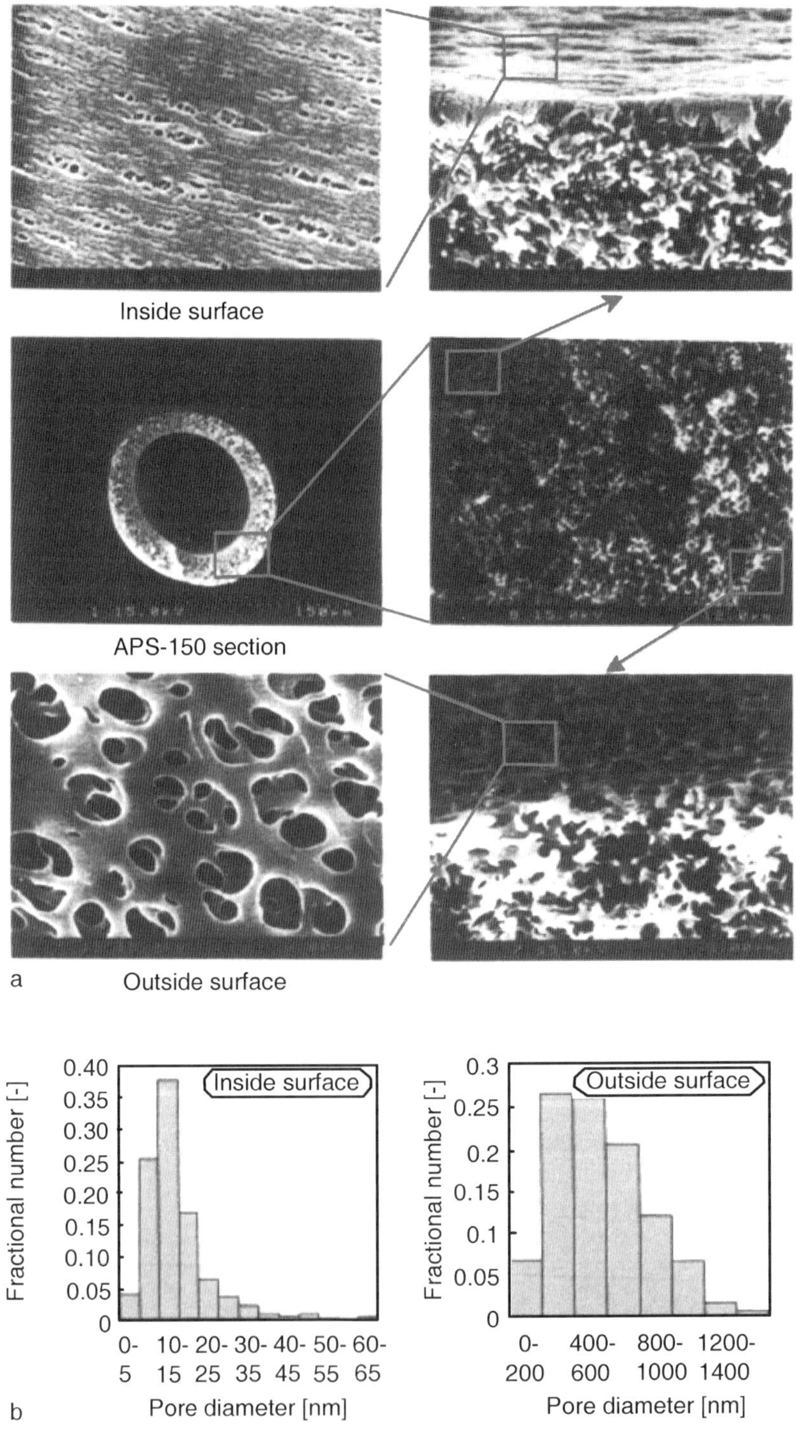
Inside surface
APS-150 section
a
Outside surface
Fractional number [-]
0.40
0.35
0.30
0.25
0.20
0.15
0.10
0.05
0
Inside surface
0-5 10-15 20-25 30-35 40-45 50-55 60-65
Pore diameter [nm]
Fractional number [-]
0.3
0.25
0.2
0.15
0.1
0.05
0
Outside surface
0-200 400-600 800-1000 1200-1400
Pore diameter [nm]
b

Fig. 5.16a,b. Field emission scanning electron microscopy of APS-150. **a** Inside surface and outside surface (scanning voltage 15 kV). **b** Distribution of pore diameter determined by FE-SEM. Reprinted from [28]. Copyright 2002, with kind permission from Elsevier

pore model (DSPM) [30, 31] are given in Table 5.9. It seems from Table 5.9 that the pore sizes of the PT series membranes do not change very much, as calculated by the DSPM model. However, the pore sizes of the PC series vary significantly according to AFM measurement. Not only do the pore sizes change significantly, but the agreement between AFM and DSPM is poor. The order of the pore size change is completely reversed. From these data, it is clear that the DSPM model could not be used for pore size determination of the membranes. But it is also clear that the pore sizes of the active layer depend on the reaction time as well as the concentration of BPA during interfacial polymerization.

Kim et al. [11] measured the pore sizes of a polysulfone membrane by using both SEM and AFM techniques. The PSf (20 wt%) solution in NMP was spread on a glass plate and immediately immersed in the nonsolvent bath consisting of pure water (sample 1) and a 20/80 mixture of water/NMP by weight (sample 2). Figure 5.1a and b shows the surface structures of the PSf membranes imaged by SEM. In the case of the pure water bath (Fig. 5.1a), very small domains exist. In the case of the mixed solvent bath (Fig. 5.1b), pores with varying sizes exist on the membrane surface. The average pore diameter is about 112 nm. Figure 5.2 shows the surface of PSf membranes imaged by AFM. Figure 5.2a (sample 1) shows a typical nodular structure with interconnected cavity channels between the agglomerated nodules. In the case of sample 2 (Fig. 5.2b), there is no nodular structure, but large pores exist throughout the membrane surface. Figure 5.19 shows the pore size distribution obtained by SEM and AFM of sample 2. The mean diameter was 112 nm for SEM and 146 nm for AFM [11].

The difference was caused most likely by the following:

1. The AFM image and the accompanying vertical displacement profiles reflect both the pores and the surrounding depressions in the membrane surface layer. In contrast, the SEM photomicrograph renders only the defects (pores) with minimal information on the surrounding surface depression due to the two-dimensional character of the image.
2. SEM usually underestimates pore diameter due to the metal coating, which is necessary to increase the conductivity and is likely to lead to reductions in the pore size. The pore diameter is varied depending on the coating rate, coating period, and pore shape. The pore shape might not necessarily be cylindrical but be funnel-shaped, so such a coating could reduce the pore size. Structural change may also occur due to the damage caused by the electron beam or by the requirement to operate in a high vacuum. Based on their observation, Kim et al. concluded that the pore diameters obtained from AFM are more accurate [11].

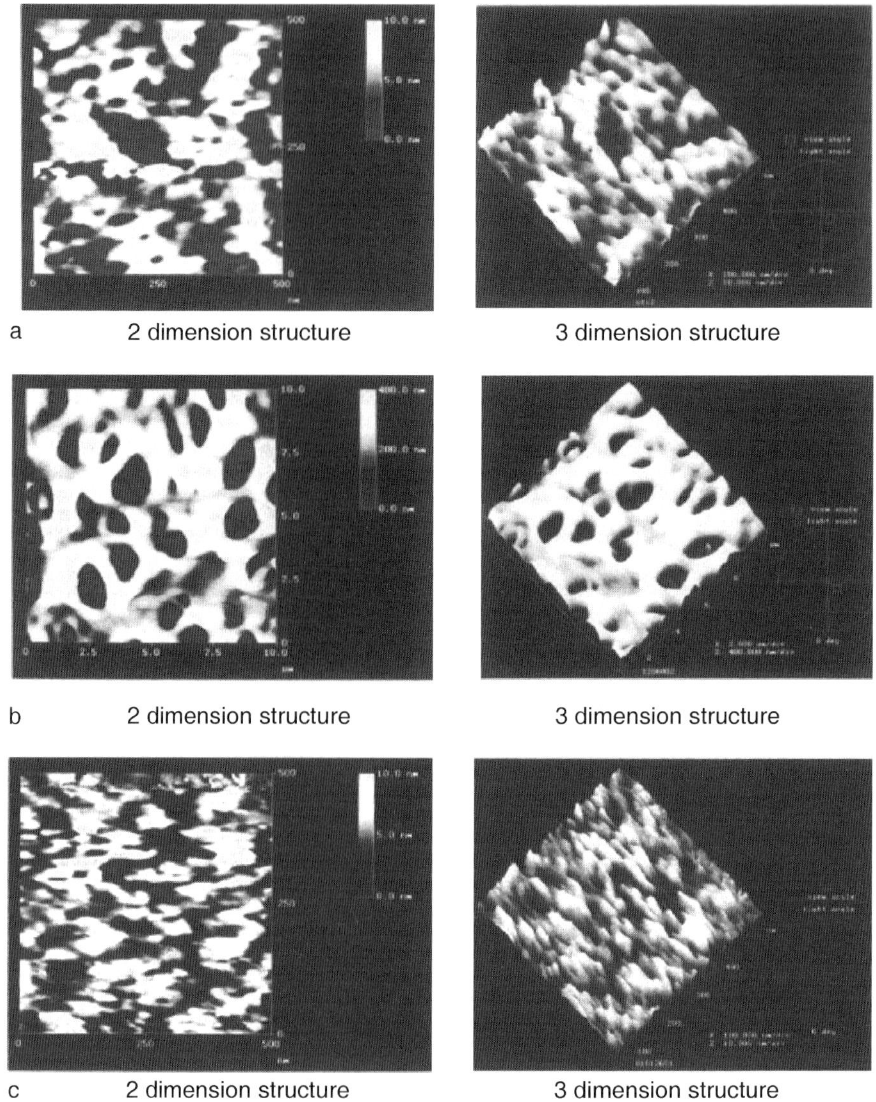

Fig. 5.17a–c. Tapping mode AFM images. **a** APS-150 inside (scan size 500 nm, scan rate 0.4265 Hz). **b** Outside (scan size 10 μm, scan rate 0.4002 Hz) with generally used silicon single-crystal probe and J-scanners. **c** APS-150 inside (scan size 500 nm, scan rate 0.3290 Hz) with highly sharpened silicon single-crystal probe and E-scanners with a smaller maximum scan area and height. Reprinted from [28]. Copyright 2002, with kind permission from Elsevier

Hernández et al. [6] measured the mean pore size of a UF membrane (N0015, Nuclepore® filters) by different techniques, and the results are given in Table 5.10. The authors further suggested that from AFM, more spectacular achievements could be obtained at nanometer range where the SEM techniques start to lose resolution.

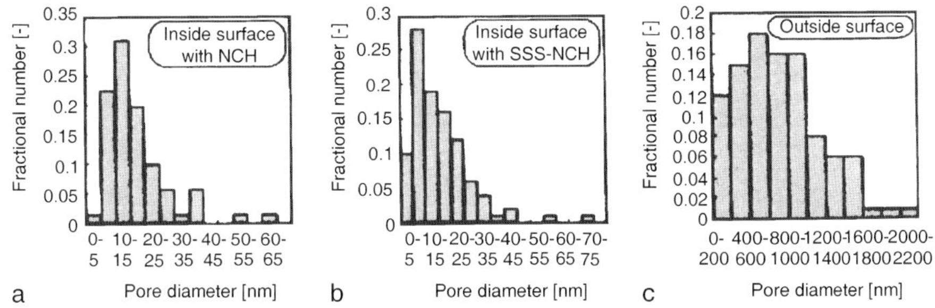

Fig. 5.18a–c. Distribution of pore diameter on the inside and outside surface: an NCH cantilever was used for **a** and **c** and an SSS-NCH cantilever was used for **b**. Reprinted from [28]. Copyright 2002, with kind permission from Elsevier

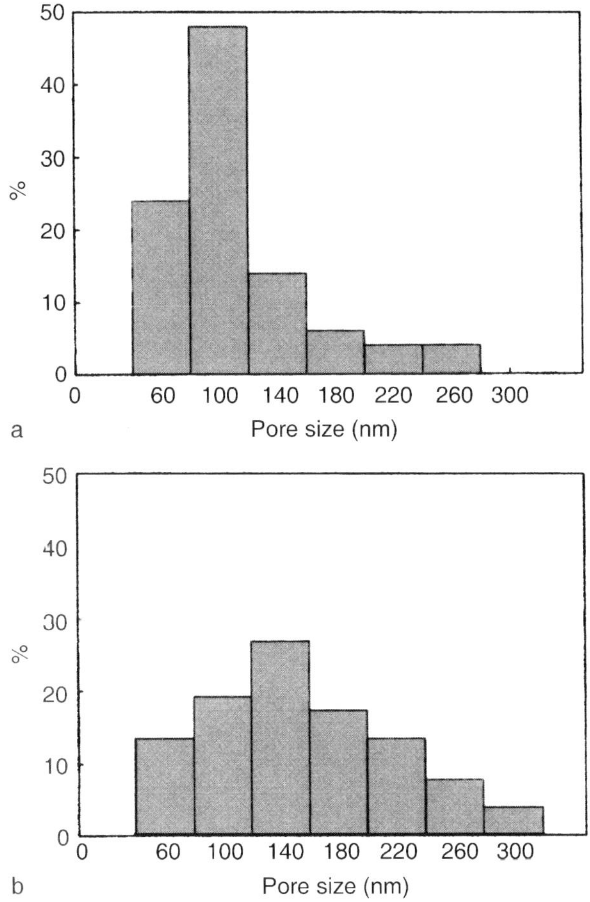

Fig. 5.19a,b. Pore size distribution of PSf membrane of sample 2. **a** Obtained from SEM. **b** Obtained from AFM. Reprinted from [11]. Copyright 1999, with kind permission from Elsevier

Table 5.10. Mean pore diameters obtained from N0015 membranes from different characterization methods. All of them fitted the log-normal distribution

Gas adsorption/desorption (nm)	Solute retention (nm)	AFM (nm)
20 ± 0.34	15.9 ± 3.4	23 ± 3

Table 5.11. Average pore size of PEI hollow fiber membranes from AFM images (inner surface) and from UF experiments

Air gap (cm)	Average pore size (nm) From AFM image (inner surface)	From UF experiment
10	30.0	16.0
30	31.0	16.5
50	32.0	17.8
70	38.0	19.0
90	41.0	20.5

Feng et al. [32] measured the pore sizes of PEI hollow fibers prepared at different air gaps by AFM and the solute transport technique. The results are given in Table 5.11. The table also shows that the pore sizes determined by AFM are always larger than those calculated from solute transport data.

In general, it has been observed that the mean pore sizes measured by AFM images are always bigger than those measured by solute transport (retention) data. Khulbe et al. [33] reported that the mean pore sizes obtained from the AFM images were twice as large as those calculated from the solute transport data. Singh et al. observed that the mean pore sizes in NF and UF membranes, when measured by the AFM technique, were about 3.5 times as large as the ones calculated from the solute transport technique [20]. Bessières et al. also observed that AFM gave 2 to 4 times larger diameters than those obtained from solute (ethylene glycol) transport [26].

In this context, it is worth noting the observation made by Kasper et al. [34]. They studied Cuprophan® as well as modified membranes containing 5, 10, 15, 20, 40, and 100% diethylaminoethylcellulose (DEAE) by AFM. The surfaces were observed in air or in a swollen state under water. In water, they reported some entrance funnels probably leading to pores.

According to Bessières et al. [26], pore sizes obtained from solute retention data correspond to a minimal size of the pore constriction experienced by the solute while passing through the pores. On the other hand, pore sizes measured by AFM correspond to the pore entrances, which are funnel-shaped and have a maximum opening at the entrances. *However, it should not be forgotten that geometry of the probe tip also plays a role in measuring the pore sizes by AFM* and so does the morphology or configuration of the membrane's surface itself.

5.2.3 Effects of Membrane Preparation and Posttreatment Parameters on Pore Size and Pore Size Distribution

It is well known that the morphology and the performance of membranes depend largely on the conditions under which the membranes are prepared. It is interesting, therefore, to review the works in which pore structures are studied by AFM for membranes that were prepared under different conditions.

Zhang et al. prepared flat cellulose membranes under different conditions, changing the temperature and *N*-methylmorpholine-*N*-oxide (NMMO) concentration of the coagulation bath and cellulose concentration of the casting solution. The solvent for making casting solution was also NMMO [35]. The codes of the membranes and the detailed information on the membrane preparation are given in Table 5.12. The surfaces of these membranes were studied by contact mode AFM. Table 5.13 shows the corresponding permeation properties, pore size statistics, and roughness parameters.

Figure 5.20 shows the AFM images of the membranes fabricated at different coagulation bath temperatures (T-20 to T-55 series), while Fig. 5.21 shows the pore size distributions of those membranes. Figure 5.22 shows the AFM images of the membranes that were fabricated at different NMMO concentrations in the coagulation bath (N-7.5 to N-30 series), while Fig. 5.23 shows the pore size distributions of those membranes. Figure 5.24 shows the AFM images of the membranes fabricated with different cellulose concentrations in the casting dope (C-9 and C-11 series). From Tables 5.12 and 5.13, it seems that the mean pore size, μ_p, increases together with the standard deviation, σ_p, as the temperature of the coagulation bath increases. This tendency (increase of σ with an increase in μ) seems common in all T, N, and C series.

Based on the AFM images, Zhang et al. further explained the following [35]:

> When a cast film is immersed in a coagulation bath, the casting solution at the surface that is in contact with the coagulation media will split into two phases, i.e. polymer-poor phase and polymer-rich phase [36]. After solidification, the polymer-poor phase will become pores, while the polymer-rich

Table 5.12. Casting solutions and the conditions of coagulation in cellulose membrane formation [35]

Membranes	Casting solution concentration (wt.%)	Temperature of coagulation bath (°C)	NMMO concentration of coagulation bath (vol.%)
T-20	7	20	0
T-35	7	35	0
T-45	7	45	0
T-55	7	55	0
N-7.5	7	25	7.5
N-15	7	25	15
N-30	7	25	30
C-9	9	25	0
C-11	11	25	0

Table 5.13. Permeation properties and surface analysis of cellulose membranes

Membranes	Permeation properties J^a (mL cm^{-2} h^{-1})	R^b (%)	Pore size (nm) $\mu_p{}^c$	σ^d	Mean roughness (nm)
T-20	0.70	41.5	99.87	25.1	2.422
T-35	1.08	14.3	106.4	29.8	2.224
T-45	1.11	< 5	145.4	35.9	2.256
T-55	4.70	< 5	175.2	46.4	7.702
N-7.5	0.64	< 5	170.3	49.6	1.600
N-15	0.89	< 5	177.0	48.7	1.807
N-30	1.27	< 5	227.0	65.6	1.908
C-9	2.68	72.4	159.6	32.2	5.253
C-11	1.08	81.6	98.70	26.6	4.174

[a] Pure water flux
[b] Rejection of BSA
[c] Mean pore size
[d] Standard deviation

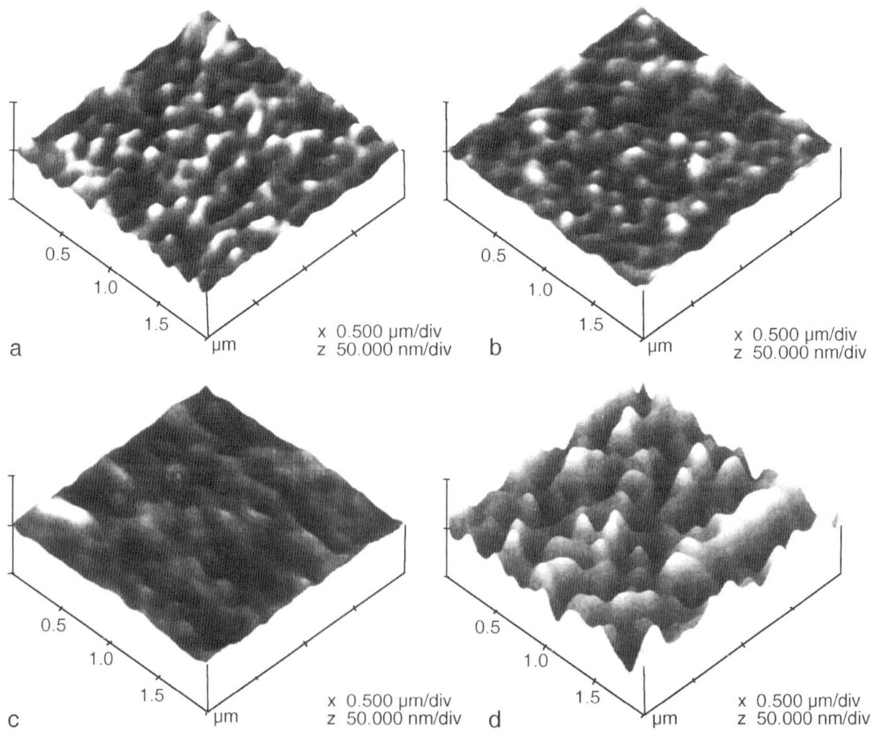

Fig. 5.20a–d. AFM images of cellulose membranes: **a** T-20, **b** T-35, **c** T-45, and **d** T-55. Reprinted from [35]. Copyright 2002, with kind permission from Wiley

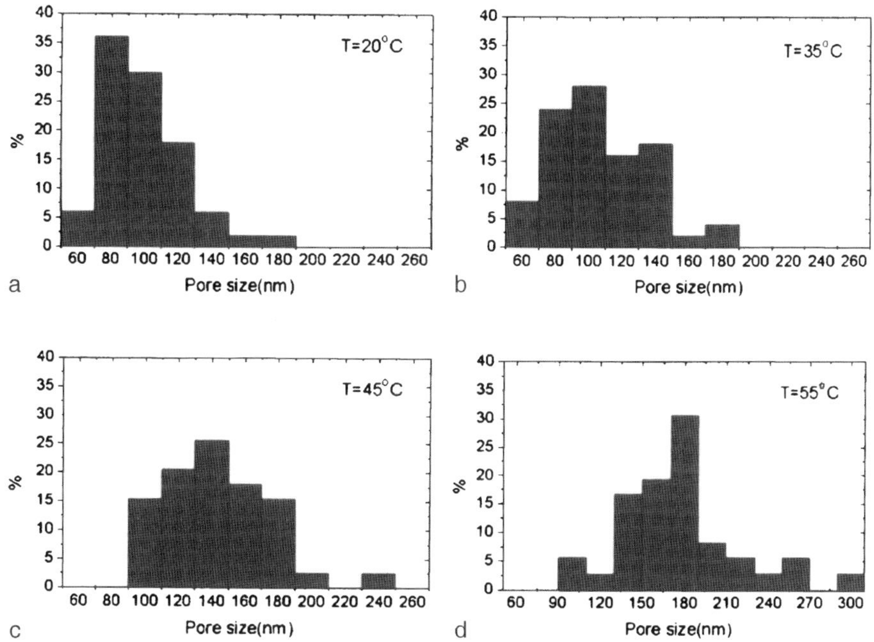

Fig. 5.21a–d. Pore size distribution of cellulose membranes: **a** T-20, **b** T-35, **c** T-45, and **d** T-55. Reprinted from [35]. Copyright 2002, with kind permission from Wiley

phase will form a polymer matrix [36]. It is believed that the solidification of cellulose is more rapid when the temperature of the coagulation bath is higher due to higher diffusion rates of NMMO and water. Accordingly, for a higher temperature of the coagulation bath, demixing occurs throughout the polymer solution instantly during the rapid solidification process. Therefore, the polymer is solidified before the merging of the polymer-rich phase, and larger pores are formed in comparison with those of the membranes prepared from the coagulation bath with a lower temperature.

Figure 5.24a shows a typical pore structure of a C-9 membrane in which large pores exist throughout the whole membrane surface. On the other hand, Fig. 5.24b shows a nodular structure with interconnected cavity channels between the agglomerated nodules for a C-11 membrane. It is believed that the cellulose concentration affects the chain entanglement. A higher cellulose concentration results in a membrane with a smaller pore size, lower flux, and higher solute rejection.

Khayet et al. prepared a number of poly(vinylidene fluoride) hollow fiber membranes for ultrafiltration using the solution spinning method. *N*,*N*-dimethylacetamide (DMAc) was the solvent, and ethylene glycol was the nonsolvent additive [37]. The effect of the concentration of ethylene glycol in the PVDF spinning solution, as well as the effect of the concentration of ethanol either in the bore liquid or in the coagulation bath, on the morphology of the hollow fiber membrane was stud-

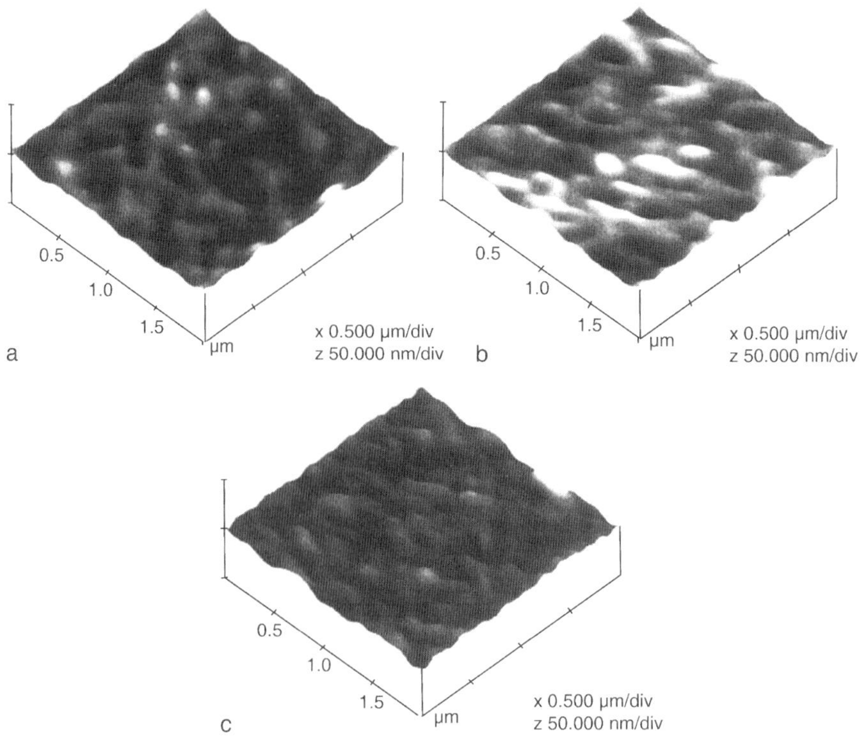

Fig. 5.22a–c. AFM images of cellulose membranes: **a** N-7.5, **b** N-15, and **c** N-30. Reprinted from [35]. Copyright 2002, with kind permission from Wiley

ied. It was reported that the pore sizes determined by AFM (TM-AFM) and by other methods all increased as the concentration of ethylene glycol in the spinning solution increased and when ethanol was added to either the bore liquid or the external coagulation bath. From AFM studies, it was concluded that the pore sizes of the inner surface (70–157 nm) were larger than those of the outer surface (50–143 nm). The mean pore size determined by AFM was about two times larger than the pore size determined by the gas permeation test (~ 2.2 for the inner pore size and ~ 1.8 for the outer pore size) and was about three times larger than that calculated from the solute transport experiments (~ 3.3 for the inner pore size and ~ 2.7 for the outer pore size). This indicates that the pore sizes of the hollow fibers depend on the physical and chemical properties of the bore fluid as well as those of the coagulation bath. Of course, there are many other factors that affect the pore structure of hollow fibers.

Bowen and Doneva measured the mean pore diameters of five Desal G-series thin film membranes (Osmonics, Vista, USA) with molecular weight cutoff values of 1000; 2500; 3500; 8000; and 10 000 [38]. The MWCO values of these membranes were determined by the manufacturer from the data obtained by UF experiments that were carried out using 1.0 g L^{-1} polyethylene glycol (PEG) solutions at an operating pressure of 827 kPa and at 25 °C. Pore sizes of these membranes were also calculated

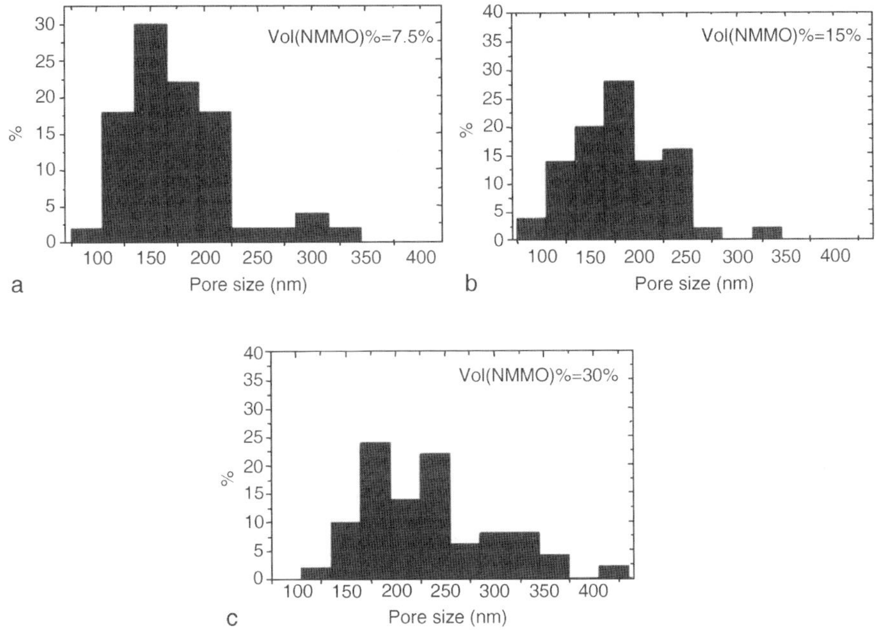

Fig. 5.23a–c. Pore size distribution of cellulose membranes obtained from AFM images: **a** N-75, **b** N-15, and **c** N-30. Reprinted from [35]. Copyright 2002, with kind permission from Wiley

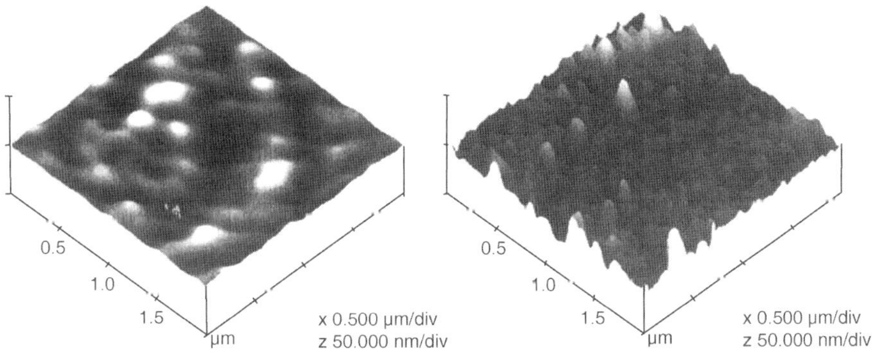

Fig. 5.24a,b. AFM images of cellulose membranes **a** C-9 and **b** C-11. Reprinted from [35]. Copyright 2002, with kind permission from Wiley

by applying the following equations to the above MWCO, as derived by Bowen and Doneva [38] and based on MWCO:

$$\lambda' = r_s / r_p \tag{5.2}$$

where r_s and r_p are the solute and pore radii (in nm), respectively, and λ' is the ratio of solute radius to pore radius. They used three different expressions for r_s. All three expressions are based on the Einstein-Stokes equation [39]. The first one is applicable

Table 5.14. Pore diameter for Desal G series membranes obtained from AFM measurements and MWCO data

Membrane	MWCO	AFM mean pore diameter (nm) [a]	Mean pore diameter from MWCO (nm)		
			Equation 3	Equation 4	Equation 5
GE	1000	1.83 (±0.35)	1.9	2.4	2.0
GH	2500	2.23 (±0.46)	2.6	3.8	3.2
GK	3500	2.23 (±0.51)	3.0	4.4	4.2
GM	8000	2.82 (±0.69)	4.4	6.4	6.8
GN	10 000	3.14 (±0.93)	4.8	7.2	7.6

[a] Values in parentheses are standard deviations

of small solutes (M_w in the range 180 – 135 Da):

$$r_s = 0.0461 M_w^{0.359} \tag{5.3}$$

where M_w is the molecular weight of solutes.

The second one is for the M_w range of 300 – 10 000 Da:

$$r_s = 0.045 M_w^{0.44} . \tag{5.4}$$

The third one is for the M_w range of 600 – 35 000 Da:

$$r_s = 0.01673 M_w^{0.557} . \tag{5.5}$$

The results are given in Table 5.14.

From Table 5.14, it seems that the pore diameters obtained from AFM measurements are less than those obtained from the MWCO data. This is opposite to the findings made by other researchers.

Feng et al. [32] studied the morphology of the inner and outer surfaces of hollow fibers fabricated from poly(etherimide) by TM-AFM. The hollow fibers were fabricated by the dry–wet phase inversion method at two different bore fluid flow rates, 0.1 and 0.4 mL min^{-1}, and their effect on the surface morphology was investigated. The average pore sizes on the inner surface were 39.8 and 81.9 nm, respectively, for 0.1 and 0.4 mL min^{-1}, while those on the outer surface were 218.4 and 93.4, respectively, for 0.1 and 0.4 mL min^{-1}. It is interesting to note that the pore size increased with an increase in the bore fluid flow rate at the inner surface, while the opposite was the case at the outer surface.

5.3 Roughness of the Membrane Surface

5.3.1 Roughness Parameters

The definitions of three roughness parameters are given in Chap. 3. All roughness parameters can be calculated from the AFM image with an AFM software program.

However, specific roughness parameters depend on the curvature on the membrane surface and also the size of the cantilever tip, as well as on the treatment of the captured images (plane fitting, flattering, etc.). Thus, the roughness parameter should not be considered as an *absolute* value that represents surface roughness.

The roughness parameter is one of the best parameters for comparing different membranes. Furthermore, it can be correlated with membrane performance and other surface properties such as pore size distribution. The ranges of roughness parameters for membranes used in different processes were discussed in the summary of Chap. 4. Bowen et al. also studied a series of UF membranes with different MWCOs and reported that the surface roughness parameter increases with an increase in pore size [10].

5.3.2 Effects of Membrane Preparation and Posttreatment Parameters on Roughness Parameters

As mentioned in Chap. 4, Khulbe et al. [40] prepared dense flat sheet poly(phenylene oxide) membranes using different solvents, including carbon disulfide, benzene, 1,1,2-trichloroethylene, toluene, chlorobenzene, and bromobenzene, for preparation of casting dopes. It was observed that the mean surface roughness of the top side (the side that was in contact with air) of the membrane was correlated to the boiling point of the solvent. On the other hand, the roughness parameters of the bottom surface (the side that was in contact with the glass plate) of the membranes did not change very much. The roughness of the bottom surface could be affected by the surface of the glass plate, as many things occur during the membrane gelation process.

It is well known that the solvent/nonsolvent exchange that takes place in the nascent membrane during the gelation process governs the membrane surface roughness. For example, Okada and Matsuura studied the macroscopic roughness of cellulose acetate membrane surfaces using aqueous solutions of different sodium chloride concentrations as gelation media in the phase inversion technique. The amplitude of the wavy pattern decreased with a decrease in water activity, while the wavelength was almost unchanged [41]. Alsari et al. [42] studied the effect of the concentration of sodium dodecyl sulfate (SDS) in the gelation media on the roughness of poly(ether sulfone) ultrafiltration membranes using TM-AFM. They reported that there was no obvious relationship between the SDS concentration and the roughness parameter of the membranes.

Ariza et al. studied the surfaces of three different types of membranes by AFM (six membranes in total). Two of them were polysulfone membranes: one was a supported ultrafiltration commercial membrane (PSf), and the other was a symmetric experimental membrane (PSC, polysulfone composite) [43]. Two were composite RO polyamide/polysulfone membranes: one was a commercial membrane supplied by FilmTec called NF45, and the other was a laboratory-made membrane called BO, both having polyamide as the active layer. The final two were experimentally activated membranes called DPA-2 and DTA-2, obtained by adding a given amount of di-2-ethylhexylphosphoric acid (DEHPA) and di-2-ethylhexylthiophosphoric acid (DTPA), respectively, to BO as carriers when the polyamide skin layer was formed by in situ polymerization. The above six membranes were studied by AFM before and after being irradiated by X-rays for periods from 15 to 180 min.

Figures 5.25, 5.26, and 5.27 show the AFM images (scan size 1×1 μm) of the fresh membranes and the membranes after 180 min of irradiation, except for BO. Table 5.15 shows the data on the average roughness parameters of fresh and 180-min-irradiated membranes obtained from AFM images of a scan size of 1×1 μm.

In Fig. 5.25, the morphologies of the PSf and PSC membranes seem different, because a nodular surface structure with an average nodule size of 40 nm is observed in PSf membranes while the PSC membrane does not show a very well-defined nodular structure. No significant differences can be observed between fresh and after 180-min-irradiation membranes, with only a slight decrease in roughness (Table 5.15). Looking into the NF45 membrane also shows little change before and after irradiation, with a slight increase in the roughness parameter after irradiation (Table 5.15). Similar to the polysulfone membranes (PSf and PSC), the commercial polyamide/polysulfone composite membrane (NF45) is featured by a nodular surface with an average nodular size of 125 μm. On the other hand, an ill-defined structure

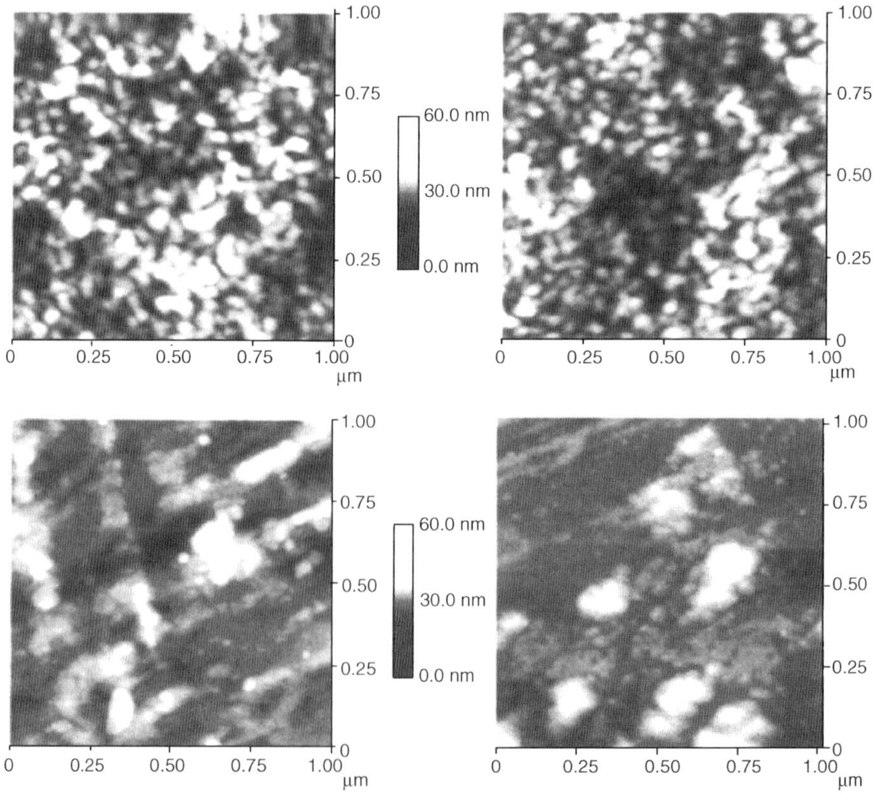

Fig. 5.25. The AFM topographic images of PSf, *top*, and PSC, *bottom*, both fresh, *left*, and after 180 min of irradiation, *right*. Reprinted from [43]. Copyright 2003, with kind permission from Wiley

Table 5.15. Average roughness, R_a, for fresh membranes (scan size 1 × 1 µm) and 180-min-irradiated membranes (scan size 1 × 1 µm)

Membrane	Fresh (nm)	After 180 min (nm)
PSf	3.2	3.1
PSC	2.5	2.0
NF45	22.7	25.8
BO	30.2	33.7
DPA-2	1.1	5.6
DTA-2	6.5	10.5

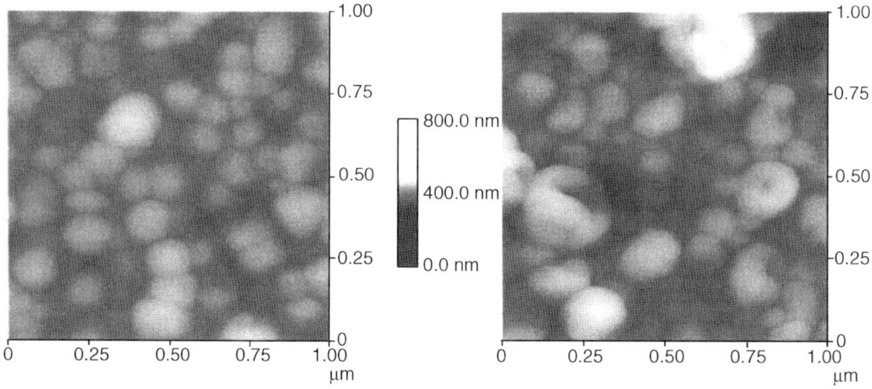

Fig. 5.26. AFM 1 µm images of commercial polyamide membrane NF-45, fresh, *left*, and after 180 min of irradiation, *right*. Reprinted from [43]. Copyright 2003, with kind permission from Wiley

was observed in the laboratory-made composite membrane (BO). The roughness parameters of the composite membranes, both commercial and laboratory-made, are an order of magnitude higher than the PSf and PSC membranes, which is attributed to fluctuation between aqueous and organic phases during the interfacial in situ polycondensation, by which the polyamide skin layer was formed [44]. Figure 5.27 shows the AFM images for DPA-2 and DTA-2 membranes before and after irradiation. The roughness parameters of the fresh DPA-2 and DTA-2 membranes are less than that of the fresh BO membrane 5.15), which suggests the influence of the addition of the carriers on the miscibility between aqueous and organic phases during the polycondensation reaction.

When the membranes were irradiated, PSf and PSC showed little changes in the roughness parameters, while the roughness parameter increased significantly for DPA-2 and DTA-2 membranes. Together with the morphological changes that occurred in the DPA-2 and DTA-2 membranes, it can be concluded that the photodegradation of polysulfone, which was detected by an independent experiment, does not affect the membrane surface morphology, but the photodegradation of the surface of the activated membranes caused the increase in their roughness parameters.

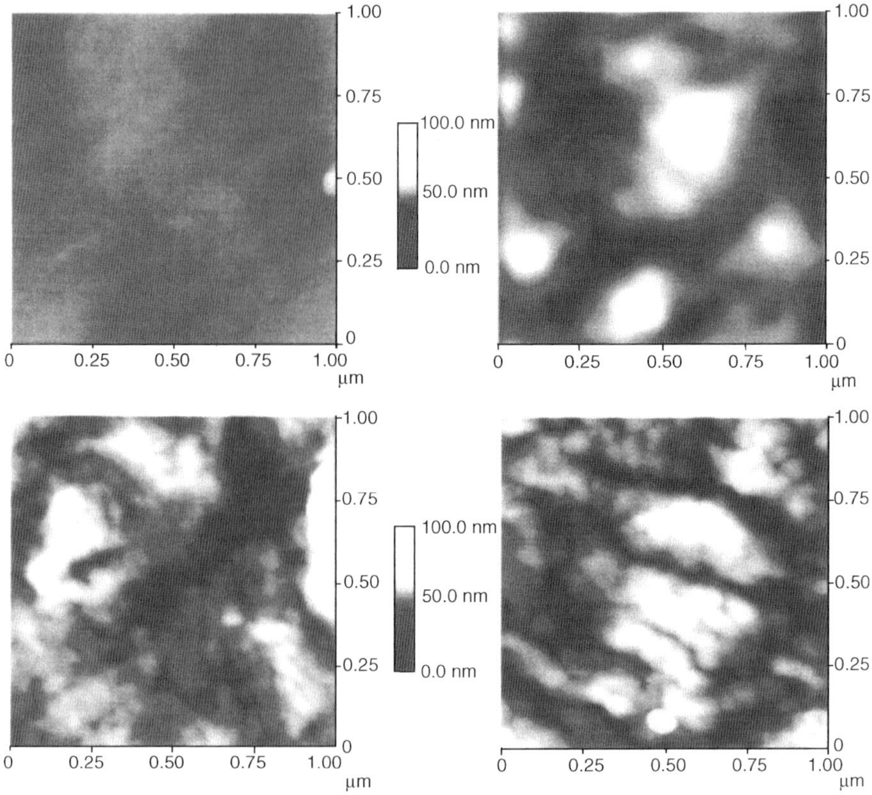

Fig. 5.27. AFM topographic images for DPA-2, *top*, and DTA-2, *bottom*, both fresh, *left*, and after 180 min of irradiation, *right*. Reprinted from [43]. Copyright 2003, with kind permission from Wiley

Barzin et al. studied hemodialysis hollow fiber membranes made of poly(ether sulfone) by UF experiments and by AFM. Hollow fibers were prepared from the solutions of PES and poly(vinylpyrrolidone) (PES/PVP 18/6 and 18/3 by weight) in N,N-dimethylacetamide by the dry–wet spinning method. The hollow fibers were then heated either in hot water (95 °C for 30 min) or in air (150 °C for 5 min) [45]. The morphology of the hollow fiber surfaces was studied by TM-AFM in terms of roughness parameters. The roughness of both the inner and outer surfaces decreased upon heat treatment, either in water or in air. The membranes heat-treated in air exhibited the lowest roughness parameters. In another study, Barzin et al. studied the separation of uric acid, vitamin B12, and bovine serum albumin (BSA) by the PES hollow fiber membranes, and their performance was correlated to the roughness of the inner surface [46].

As mentioned in Chap. 4, Khulbe et al. studied dense poly(phenylene oxide) membranes cast from solutions of different solvents by AFM [40]. PPO-CS_2 represents PPO membrane cast from PPO solution in carbon disulfide solvent, while PPO-TCE represents PPO membrane cast from PPO solution in trichloroethylene solvent.

These membranes were plasma-etched for different periods before being subjected to TM-AFM study. Figure 5.28 shows the change in the roughness parameter with etching time. In the PPO-CS_2 membrane, the roughness parameter increases up to 800 s and then levels off, while in the PPO-TCE membrane, the roughness parameter increases sharply after a lag time of about 800 s.

Kawakami et al. prepared dense and asymmetric membranes from 2,2′-bis(3,4-dicarboxyphenyl)hexafluoropropane dianhydride (6FDA) and bis[4-(4-aminophenoxy)phenyl]sulfone (APPS) by solvent evaporation (dense) and by the dry–wet phase inversion technique [47]. The surface morphology was studied by AFM. They reported that the solvent evaporation method adopted for the preparation of the dense membrane influenced the formation of nodules, while the dry–wet process in which solvent/nonsolvent exchange was involved determined the roughness of the skin layer.

Teng et al. investigated the surface property of AAm-p-aramide membranes, aromatic polyamide membranes on which acrylamide (AAm) was grafted by plasma polymerization, by AFM [48]. The membranes were prepared for pervaporation of water/ethanol mixtures. The degree of grafting AAm onto aramide membranes was calculated according to the following equation:

$$Degree\ of\ grafting\ (\%) = [(A - B)/B] \times 100 \qquad (5.6)$$

where A is the weight of the grafted aramide membrane and B is the weight of the aramide membrane before grafting.

Figure 5.29 shows the AFM image of an aramide membrane without surface modification, while Fig. 5.30 shows the AFM image of the surface of an AAm-p-aramide membrane with a degree of grafting of 8.2%. The change in surface morphology depended on the degree of grafting. Table 5.16 shows that the roughness parameters increase with an increase in degree of grafting.

Espinoza-Gómez and Lin prepared negatively charged hydrophilic UF membranes from the blend of acrylonitrile-vinyl acetate (CP16) and acrylonitrile-vinyl acetate-sodium p-sulfophenyl methallyl ether (CP24) [49]. The compositions of membrane casting solutions for nine different membranes are summarized in Table 5.17.

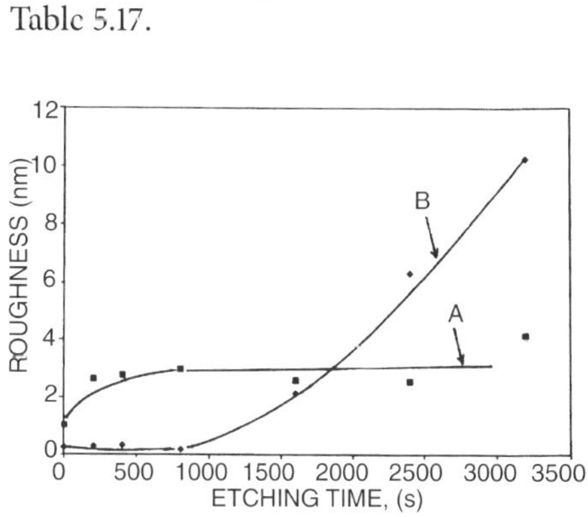

Fig. 5.28. Roughness parameter (nm) of the etched membrane (scan range 1 μm) against etching time. Line A indicates PPO-CS_2 and line B indicates PPO-TCE. Reprinted from [Chap. 4, 54]. Copyright 2000, with kind permission from Elsevier

Table 5.16. Surface roughness values of membranes measured by AFM

Membrane	Degree of grafting (%)	R_a (nm)	R_q (nm)	R_z (nm)
Unmodified aramide	0	0.273	0.345	0.556
AAm-p-aramide	6.1	0.511	0.666	1.092
AAm-p-aramide	8.2	0.789	1.064	2.989
AAm-p-aramide	21	0.959	1.215	3.486

R_a average plane roughness, R_q square mean plane roughness, R_z 10-point mean plane roughness

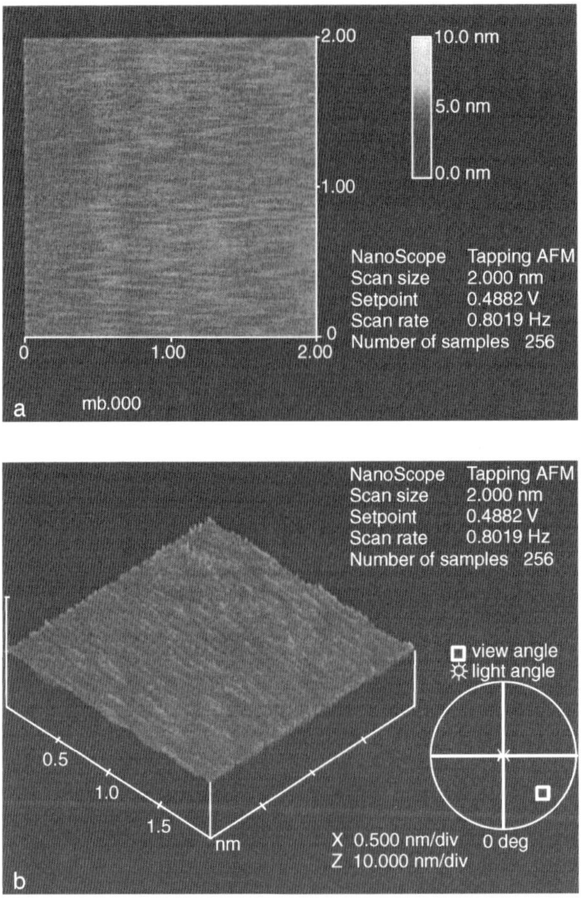

Fig. 5.29a,b. AFM images of the surface of the unmodified aramide membrane. **a** Two-dimensional image. **b** Three-dimensional image. Reprinted from [48]. Copyright 2000, with kind permission from Elsevier

The surface morphology of the wet membranes was characterized by contact mode AFM (CT-AFM). The AFM images of all the membranes are shown in Fig. 5.31. The figure indicates that the surface roughness of the membranes depends on the composition of the membrane casting solution. Figure 5.32 shows the average roughness as functions of total solid concentration (wt.%) and CP24 wt.% in the casting solution. It is clear that the surface roughness decreases as the total polymer concen-

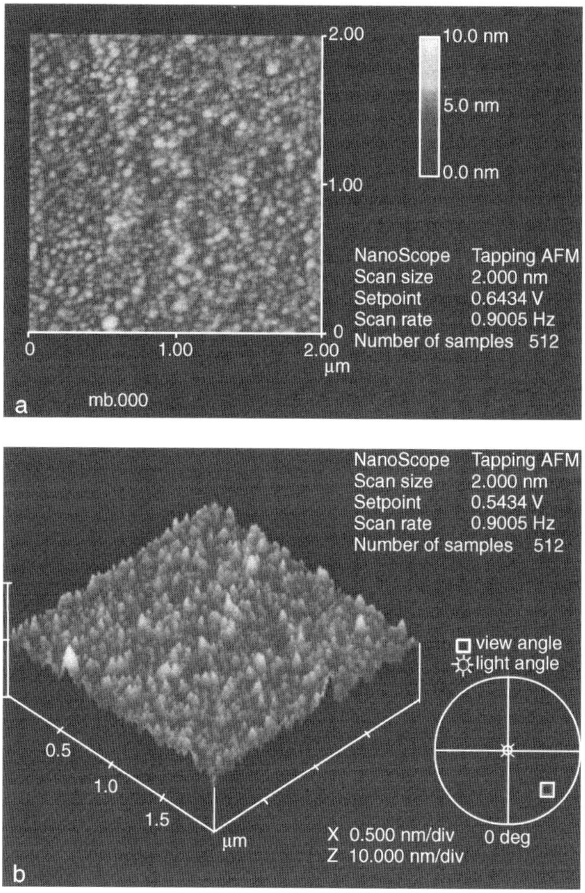

Fig. 5.30a,b. AFM images of the surface of the AAm-p-aramide (8.2 wt.%) membrane. **a** Two-dimensional image. **b** Three-dimensional image. Reprinted from [48]. Copyright 2000, with kind permission from Elsevier

tration increases and, for a given total polymer weight percentage, surface roughness also decreases as the CP24 weight percentage increases.

As mentioned in Chap. 4, Finot et al. studied the surface of plasma-coated HMDSO (hexamethyldisiloxane) membranes posttreated by CF_4/Ar plasma [50]. CF_4/Ar plasma produces F· and CF_x· radicals, which first etch the surface of the deposited layer, leading to the exodiffusion of reducing agents (H_2, CH_x). The roughness parameters and the radii of the curvature of the nodule aggregates of the membranes are given in Table 5.18. Together with Tables 4.5 and 4.6, it seems that the roughness parameters increase as the nodule or nodule aggregate size increases. The surface becomes smoother on treating the surface by CF_4/Ar plasma.

Wang et al. [51] studied the separation of a water/acidic mixture by pervaporation using a plasma-treated asymmetric poly(4-methyl-1-pentene) (TPX) membrane, which was further dip-coated with polyacrylic acid (PAA). The asymmetric TPX membrane was prepared by the wet phase inversion technique. Membranes were treated with residual air plasma in a tubular-type reactor. The modification of the

Table 5.17. Composition of the CP16/CP24 membrane casting solutions

	Increasing the contents of solids at casting solution		
Increasing the negative load of the membrane surface	HE 1 CP24 = 0% CP16 = 12% LiBr = 4% NMP = 84%	HE 4 CP24 = 0% CP16 = 14% LiBr = 4% NMP = 82%	HE 7 CP24 = 0% CP16 = 18% LiBr = 4% NMP = 78%
	HE 2 CP24 = 6% CP16 = 6% LiBr = 4% NMP = 84%	HE 5 CP24 = 7% CP16 = 7% LiBr = 4% NMP = 82%	HE 8 CP24 = 9% CP16 = 9% LiBr = 4% NMP = 78%
	HE 3 CP24 = 12% CP16 = 0% LiBr = 4% NMP = 84%	HE 6 CP24 = 14% CP16 = 0% LiBr = 4% NMP = 82%	HE 9 CP24 = 14% CP16 = 0% LiBr = 4% NMP = 78%

Table 5.18. Roughness parameters and curvature radii of nodule aggregates (parameters were calculated on a 2 × 2 μm area scan) on plasma-treated cellulose ester membranes

Samples	R_z (nm)	R_a (nm)	R_q (nm)	Curvature radius (nm)
S	36	14	14	620
SF	23	12	8.2	220
M	24	7.3	9.3	600
MF	17	6.8	8.5	430
H	17	5.7	7.3	150
HF	13	4.9	6.3	220

asymmetric TPX membrane surface by plasma led to a remarkable effect in the water contact angle, adhesion work, and surface roughness. Their work indicated that the plasma treatments were effective in rendering the asymmetric TPX membrane hydrophilic. The surface properties of the untreated and plasma-treated TPX membranes were studied by AFM. Further, the researchers dip-coated the plasma-treated TPX membrane with polyacrylic acid (PAA) to form a composite PAA/TPX membrane for pervaporation experiments. No peeling occurred between the asymmetric TPX sublayer and the polyacrylic acid top layer. Thus, a PAA/TPX composite membrane with a high stability and peel strength was prepared by using a plasma pretreatment procedure. This study shows the importance of surface roughness of the support in making composite membranes. The AFM results are summarized in Table 5.19.

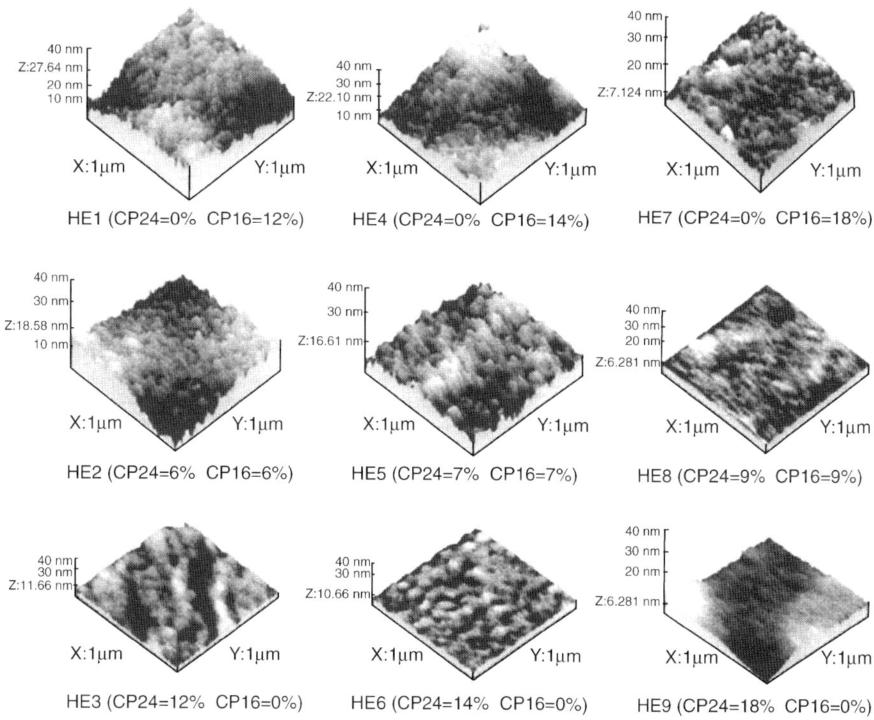

Fig. 5.31. Atomic force micrographs of CP24/CP16 negatively charged membranes. Reprinted from [49]. Copyright 2001, with kind permission from Springer-Verlag

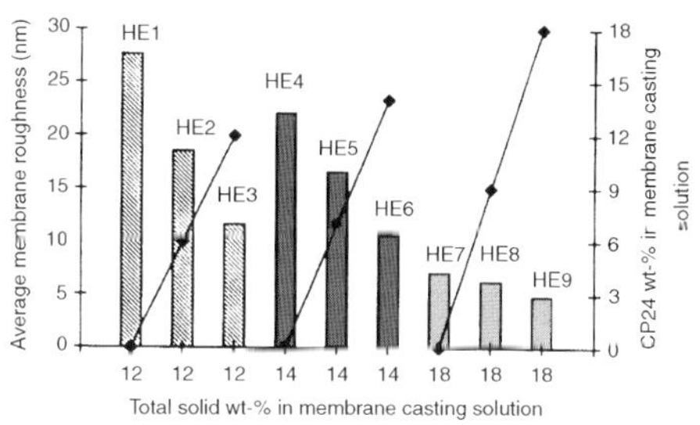

Fig. 5.32. Dependence of average membrane surface roughness on total solid content and CP24 wt.% in membrane casting solution. Reprinted from [49]. Copyright 2001, with kind permission from Springer-Verlag

Table 5.19. Roughness parameters of the surface of untreated and plasma-treated asymmetric TPX membranes

Asymmetric TPX membrane	Roughness (nm)		
	R_q	R_a	R_z
Without plasma treatment	38.5	28.6	308.1
With plasma treatment [a]	95.5	65.2	784.2

[a] Plasma treatment conditions: 5W / 30s

R_q square plane roughness, R_a average plane roughness, R_z 10-points plane roughness

5.4 Summary

Atomic force microscopy is an effective tool to investigate the surface structure of membranes and to measure pore size, pore size distribution, and roughness parameters. Moreover, membrane samples can be studied in air or under liquid. AFM usually gives pore sizes larger than those obtained by conventional methods, such as the scanning electron microscope or methods involving gas permeation and solute rejection data. This is probably because the pore sizes measured by AFM correspond to the sizes of pore entrances; if the pores are funnel-shaped, they have the maximum opening at the entrance. Pore size distributions calculated from AFM images are remarkably fitted to the log-normal distribution curve. It has also been found by AFM study that membrane morphology depends largely on the conditions of membrane preparation.

Roughness parameters obtained from AFM can also be used for the comparison of different membrane surfaces. The membrane posttreatment alters the surface roughness. By etching the surface, roughness increases, and by heating, the roughness decreases.

References

1. Kamide K, Manabe S (1980) Characterization technique of straight through porous membrane. In: Cooper AR (ed) Ultrafiltration membranes and applications. Plenum, Dordrecht, p 173; also (1981) Polym J 13:459
2. Kamide K, Manabe S (1985) Role of microphase separation phenomena in the formation of porous polymeric membranes. In: Lloyd DR (ed) Materials science of synthetic membranes. ACS Symposium Series 269. American Chemical Society, Washington, DC, p 197
3. Raman LP, Cheryan M, Rajagopalan N (1994) Chem Eng Prog 90:68
4. Smolders CA, Vugteveen E (1985) New characterization methods for asymmetric ultrafiltration membranes. In: Lloyd DR (ed) Materials science of synthetic membranes. ACS Symposium Series 269. American Chemical Society, Washington, DC, p 327
5. Zeman L, Tkacik G (1985) Pore volume distribution in ultrafiltration membranes. In: Lloyd DR (ed) Materials science of synthetic membranes. ACS Symposium Series 269. American Chemical Society, Washington, DC, p 339
6. Hernández A, Calvo JJ, Prádanos P, Tejerina F (1998) Colloids Surf A 138:391
7. Calvo JJ, Prádanos P, Hernández A, Bowen WR, Hilal N, Lovitt RW, Williams PM (1997) J Membr Sci 128:7
8. Fritzsche AK, Arevalo AR, Moore MD, Weber CJ, Elings VB, Kjoller K, Wu MC (1992) J Membr Sci 68:65
9. Dietz P, Hanssma PK, Inacker O, Lehmann HD, Hermann KH (1992) J Membr Sci 65:101
10. Bowen WR, Hilal H, Lovitt RW, Wright CJ (1999) In: Sorensen TS (ed) Surface chemistry and electrochemistry of membranes. Surfactant Science Series, Vol 79. Dekker, New York, p 1
11. Kim JY, Lee HK, Kim SC (1999) J Membr Sci 163:159
12. Ochoa NA, Pradanos P, Palacio L, Marchese J, Hernandez A (2001) J Membr Sci 187:227
13. Hansma PK, Elings VB, Marti O, Bracker CE (1998) Science 242:209
14. Albrecht TR, Quate CF (1987) J Appl Phys 62:2599
15. Marti O, Drake B, Gould SAC, Hansma PK (1988) J Vac Sci Technol A 6:287
16. Paredes JI, Martinez-Alonso A, Tascon JMD (2003) Microporous Mesoporous Mater 65:126
17. Dietz P, Hansma PK, Hermann KH, Inacker O, Lehmann HD (1991) Ultramicroscopy 35:155
18. Bowen WR, Hilal N, Lovitt RW, Williams PM (1996) J Membr Sci 110:229
19. Bowen WR, Hilal N, Lovitt RW, Williams PM (1996) J Membr Sci 110:233
20. Singh S, Khulbe KC, Matsuura T, Ramamurthy P (1998) J Membr Sci 142:111
21. Bowen WR, Hilal N, Lovitt RW, Sharif AO, Williams PM (1997) J Membr Sci 126:77

22. Bowen WR, Doneva TA (2000) J Membr Sci 171:141
23. Bowen WR, Hilal N, Lovitt RW, Williams PM (1996) J Colloid Interface Sci 180:350
24. Hilal N, Mohammad AW, Atkin B, Darwish NA (2003) Desalination 157:137
25. Hilal N, Kochkodan V, Al-Khatib L, Busca G (2002) Surf Interface Anal 33:672
26. Bessières A, Miereles M, Coratger R, Beauvillainand J, Sanchez VJ (1996) J Membr Sci 109:271
27. Gordano A, Arcella V, Drioli E (2004) Desalination 163:127
28. Hayama M, Kohori F, Sakai K (2002) J Membr Sci 197:243
29. Mohammad AW, Hilal N, Seman MNA (2003) Desalination 158:73
30. Mohammad AW (2002) Sep Sci Technol 37:1009
31. Bowen WR, Mohammad AW, Hilal N (1997) J Membr Sci 126:91
32. Feng CY, Khulbe KC, Chowdhury G, Matsuura T, Sapkal VC (2001) J Membr Sci 189:193
33. Khulbe KC, Feng CY, Hamad F, Matsuura T, Khayet M (2004) J Membr Sci 245:191
34. Kasper K, Hermann KH, Dietz P, Hansma PK, Inacker O, Lehmann HD, Rintelen Th (1992) Ultramicroscopy 42–44:1181
35. Zhang Y, Shao H, Hu X (2002) J Appl Polym Sci 86:3389
36. Cheng LP, Soh YS, Dwan A, Gryte CC (1994) J Polym Sci Part B Polym Phys 32:1413
37. Khayet M, Feng CY, Khulbe KC, Matsuura T (2002) Polymer 43:3879
38. Bowen WR, Doneva TA (2000) Surf Interface Anal 29:544
39. Bowen WR, Mohammad AW (1998) Trans IChemE 76:885
40. Khulbe KC, Matsuura T, Lamarche G, Kim HJ (1997) J Membr Sci 135:211
41. Okada T, Matsuura T (1988) Ind Eng Chem Res 27:1335
42. Alsari AM, Khulbe KC, Matsuura T (2001) J Membr Sci 188:279
43. Ariza MJ, Prádanos P, Rico R, Rodàiguez-Castellón E, Benavente J (2003) Surf Interface Anal 35:360
44. Kwak SY, Jung SC, Kim SH (2001) Environ Sci Technol 35:4334
45. Barzin J, Feng C, Khulbe KC, Matsuura T, Madaeni SS, Mirzadeh H (2004) J Membr Sci 237:77
46. Barzin J, Feng C, Khulbe KC, Matsuura T, Madaeni SS, Mirzadeh H (2004) Unpublished results
47. Kawakami H, Mikawa M, Nagaoka S (1997) J Membr Sci 137:241
48. Teng MY, Lee KR, Liaw DJ, Lin YS, Lai JY (2000) Eur Polym J 36:663
49. Espinoza-Gómez H, Lin SW (2001) Polym Bull 47:297
50. Finot E, Roualdes S, Kirchner M, Rouessac V, Berjoan R, Durand J, Goudonnet JP, Cot L (2002) Appl Surf Sci 187:326
51. Wang YC, Li CL, Chang PF, Fan SC, Lee KR, Lai JY (2002) J Membr Sci 208:3

6 Cross-sectional AFM Image

6.1 Introduction

A large number of SEM pictures have been taken to show the cross-sectional structure of integrally skinned asymmetric membranes and thin film composite membranes. Figures 6.1 [1] and 6.2 [2] are typical examples of such pictures, showing the top skin layer supported by a porous sublayer comprised of either sponge-like pores (Fig. 6.1) or finger-like pores (Fig. 6.2). Although these pictures can clearly distinguish the top skin layer from the support layer, whereby the measurement of the thickness of the top skin layer is enabled, they do not reveal the structure inside the top skin layer. Since the performance of the membrane is primarily governed by the top skin layer, whose structure will obviously control membrane performance such as flux and selectivity, it would be beneficial to have a closer view of the top skin layer.

The cross-sectional pictures taken by Panar et al. [3] (see Chap. 4) revealing the nodular structure near the top edge of the membranes are probably the first attempt to examine the detailed structure inside the top skin layer. Similar attempts have been recorded since then in the literature, but the number is limited.

There are many methods of cutting, slicing, and fractioning by using a razor blade, microtome, or ultramicrotome. Cutting can also be done at normal temperature or in liquid nitrogen. Usually hollow fibers or membranes are fractured at the liquid nitrogen temperature to keep the originality of the required testing area in cross-sectional imaging by SEM or TEM. It is, however, difficult to obtain a smooth cut of the cross section of the hollow fiber or membrane.

Cross-sectional pictures have not been taken by AFM for a long time due to difficulties involved in the preparation of a smooth cross-sectional area of a reasonable size when the membrane is sliced or fractured. An attempt to overcome these difficulties was made recently by Khulbe [4], the results of which will be briefly outlined.

6.2 Cross-sectional Images

6.2.1 Cross-sectional Images of Membranes by SEM

Kesting and Fritzsche [5] studied the cross-sectional structure of polysulfone hollow fibers they fabricated using Lewis acid/base complexes as solvents for the spinning solutions. Figures 6.3 and 6.4 are the SEM pictures of the outer edge of the hollow

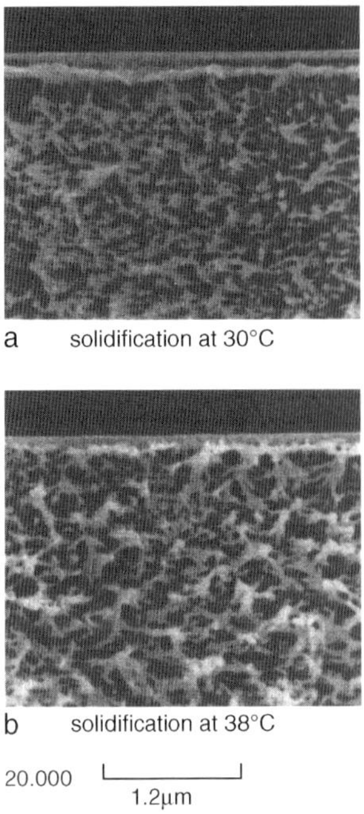

Fig. 6.1. FE-SEM photographs of cross sections of asymmetric 6FDA-BAAF membranes quenched in water at 30 °C and at 38 °C. Reprinted from [1]. Copyright 1999, with kind permission from the American Chemical Society

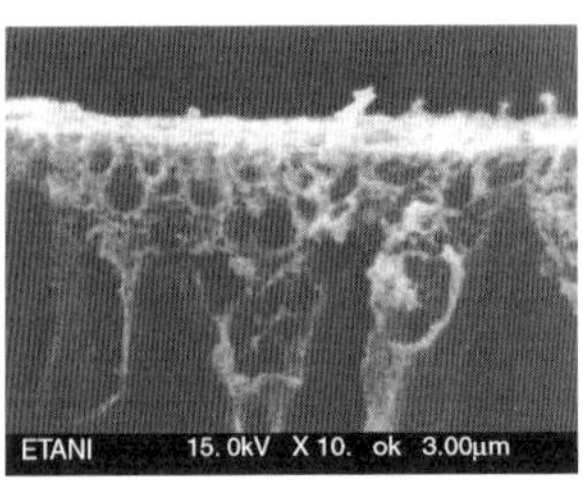

Fig. 6.2. FE-SEM image of integrally skinned asymmetric polysulfone membrane dried under T2E (ethanol). Reprinted from [2]. Copyright 1999, with kind permission from the American Chemical Society

fibers spun from a formamide/NMP solvent and a propionic acid/NMP solvent, respectively. The hollow fiber spun from formamide/NMP reveals a thick, dense skin of ~ 400 nm. The skin is composed of nodule aggregates so tightly packed at the surface that their boundaries are not discernible. Beneath this layer, the individual nodule aggregates become more readily identifiable but remain in tightly packed assembly. The skin layer of the hollow fiber spun from propionic acid/NMP is thinner than the skin of the former hollow fiber. Again, beneath the tightly packed nodule aggregates of the skin layer, less tightly packed nodule assemblies can be observed. Kesting and Fritzsche correlated the tightness of the packing of the nodular aggregates to the stability of the Lewis acid/base complex. The packing becomes less tight and the skin layer becomes thinner as the complex becomes less stable, leading to enhancement of gas flux. Kesting and Fritzsche [5] were successful in producing a high-flux polysulfone substrate to be used for a gas separation membrane after silicone rubber coating. Their gas separation membrane is known as *Prism alpha* hollow fiber. Thus, Kesting

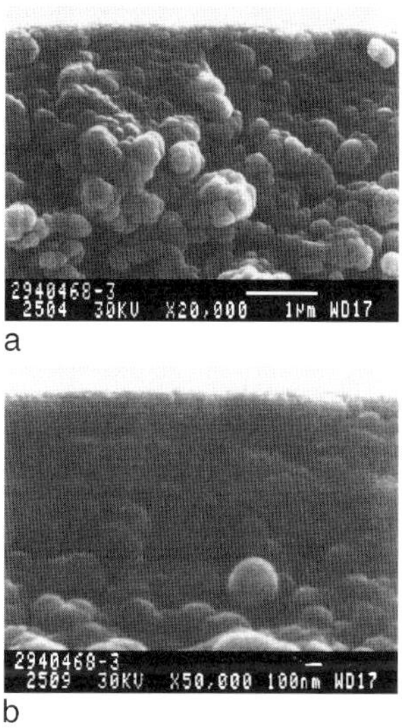

Fig. 6.3a,b. SEM photomicrograph of outer edge of the cross section of a hollow fiber membrane spun from formamide/NMP at **a** ×20 000 and **b** ×50 000. Reprinted from [5]. Copyright 1993, with kind permission from Wiley

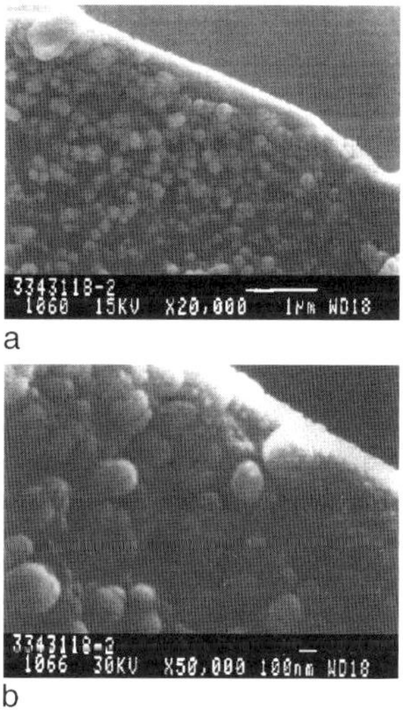

Fig. 6.4a,b. SEM photomicrograph of outer edge of the cross section of a hollow fiber membrane spun from propionic acid/NMP at **a** ×20 000 and **b** ×50 000. Reprinted from [5]. Copyright 1993, with kind permission from Wiley

and Fritzsche [5] used the information on the cross-sectional structure effectively to manufacture a high-flux gas separation membrane.

Wienk et al. [6] studied the cross section of PES/PVP hollow fiber membranes. Figure 6.5 shows the SEM pictures of the cross-section near the inner surface (a, bore side) and near the outer surface (b). A comparison of Fig. 6.5a and b indicates that the size of the nodules is smallest at the inner surface, which is the active surface, and increases in the depth direction until it gradually transforms into the porous structure near the outer surface.

Boom et al. [7] studied a PES/PVP (PVP as an additive) hollow fiber membrane by electron microscope and also observed a nodular structure. On the active surface,

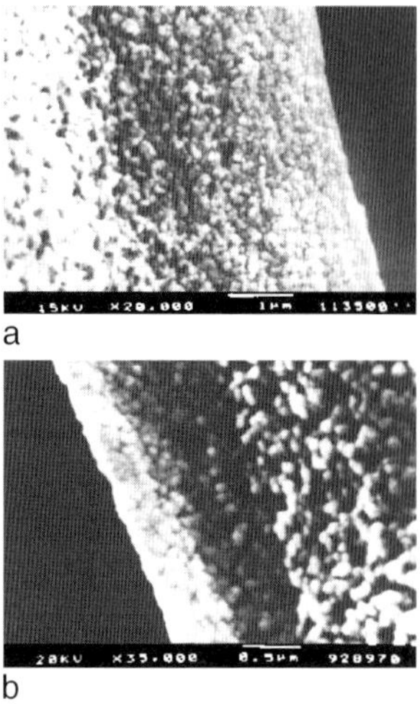

Fig. 6.5a,b. Cross-sectional pictures of PES/PVP hollow fiber **a** near the inner (bore side) surface and **b** near the outer surface. Reprinted from [6]. Copyright 1994, with kind permission from Wiley

the nodule size was estimated to be 50 nm. The nodule size increased from 50 nm (at 0.5 μm distance from the surface) to 100nm (at 5 μm below the surface). The formation of the nodular structure was ascribed to the spinodal demixing.

The Kawakami group [8] studied asymmetric polyimide membranes prepared from 6FDA[2,2′-bis(3,4-dicarboxyphenyl)hexafluoropropane dianhydride] and APPS {bis[4-(4-aminophenoxy)phenyl]sulfone} in a flat sheet to be used for gas separation and observed a typical structure of the cross section. Figure 6.6 shows the cross section of the asymmetric 6FDA-APPS membranes prepared after an evaporation time of 15, 120, and 600 s. All the structures consisted of a skin layer and a porous substructure. Figure 6.7 shows the outermost region of the asymmetric 6FDA-APPS membranes. The region of the membrane coagulated after an evaporation time of 15 s showed an ultrathin skin layer (Fig. 6.7a) and sponge-like structure characterized by the presence of macrovoids (Fig. 6.6a). On the other hand, the cross section of the membranes made after evaporation times of 120 and 600 s indicated only a sponge-like structure (Fig. 6.6b and c). Pinnau and Koros [9,10] also reported that the outermost region of the asymmetric membrane generated during the prolonged evaporation period showed a significantly different structure compared to that coagulated after a shorter evaporation time. The cross section of the membrane prepared from a shorter evaporation time indicated a sponge-like structure containing macrovoids. However, those prepared with a longer evaporation time had only sponge-like structures.

Xu et al. [11] studied the effects of both N,N-dimethylacetamide (DMAc) as a solvent additive in an internal coagulant (water) and acetic acid as a nonsolvent additive in a dope solution (PEI in DMAc) on the morphology and performance of poly(etherimide) (PEI) hollow fiber membranes for UF. Cross-sectional pictures were taken by SEM. The authors observed nodular structures on the inner and outer edges of the cross section when the amount of the acetic acid in the dope solution was increased. Pure water permeation flux increased when the nodules appeared.

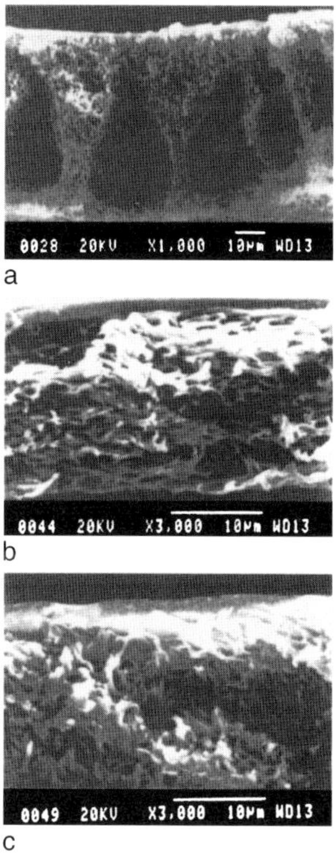

Fig. 6.6a–c. Cross section of asymmetric 6FDA-APPS membranes according to evaporation time: **a** 15, **b** 120, and **c** 600 s. Reprinted from [8]. Copyright 1997, with kind permission from Elsevier

Fusaoka et al. [12] observed a structure composed of fused and deformed polymeric spheres using FE-SEM in a cross section of the membranes, cast and dried from solutions of poly(1-trimethyl-silyl-1-propyne), poly(4-methyl-1-pentene), and some other polymers.

Fujii et al. [13] studied morphological structures of the cross section of various hollow fibers and flat sheet membranes by high-resolution field emission scanning electron microscopy. Figure 6.8 shows a cross-sectional structure of a flat sheet cellulose acetate RO membrane. The layer near the top surface is composed of a densely packed monolayer of polymeric spheres, which is supported by a layer formed with completely packed spheres. The contours of the spheres in the top layer can be observed. The middle layer is also composed of loosely packed and partly fused spheres, which are larger than the spheres in the surface layer. In the middle layer, there are many microvoids, the sizes of which are the same as the spheres. The layer near the bottom is denser than the middle layer, and the spheres are deformed and fused. Interstitial void spaces between the spheres, which may be called microvoids, are clearly observed. This structure seems common for the flat sheet as well as the hollow fiber membranes. For example, Fig. 6.9 shows a cross section of a hollow fiber made of PMMA B-2 (a copolymer containing methyl methacrylate and a small amount of sulfonate groups). The inside surface layer is composed of the dense structure of compactly packed fine polymeric particles. The particle structure of the middle layer

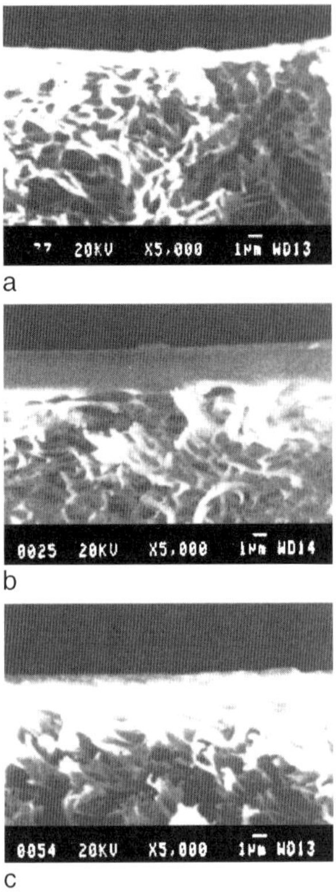

Fig. 6.7a–c. Cross section of asymmetric 6FDA-APPS membranes, outermost region, according to evaporation time: **a** 15, **b** 120, and **c** 600 s. Reprinted from [8]. Copyright 1997, with kind permission from Elsevier

of the wall is clearer than the inside surface. The outside surface layer is composed of densely packed and fused polymer spheres, and crevices between the clusters of the polymer spheres can be observed. On the outer surface of the hollow fiber, the macromolecular spheres are aligned to the direction of the fiber axis. Fujii et al. [13] also reported a change in the diameter of the macromolecular spheres from the inside surface to the middle layer and to the outside surface of the hollow fiber. Sometimes the diameters decreased, and sometimes they increased from the inside surface to the outside surface, and prediction was difficult.

The conclusions obtained by Fujii et al. [13] are as follows:

1. Membranes prepared by the dry–wet phase inversion method from glassy polymers are composed of polymeric spheres.
2. The interstitial void spaces are connected to the openings at the membrane surfaces, forming capillary pores in the membranes.
3. The pore radii are almost the same as the minimum pore radii calculated from the spherical diameters.

The authors suggest that mass transport takes place through such pores.

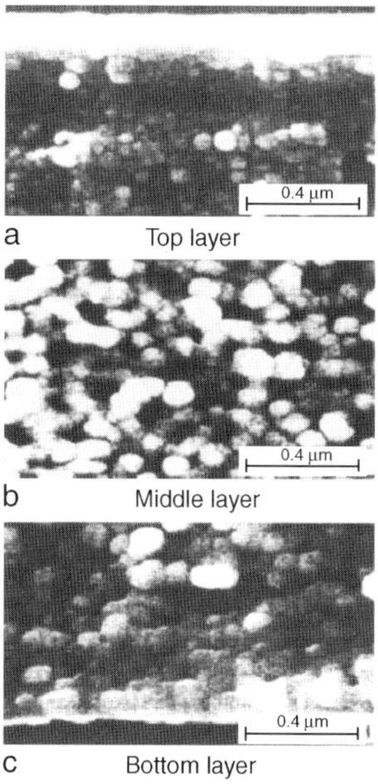

Fig. 6.8. Cross-sectional structure of CA reverse osmosis membrane. Reprinted from [13]. Copyright 1992, with kind permission from the Society of Polymer Science, Japan

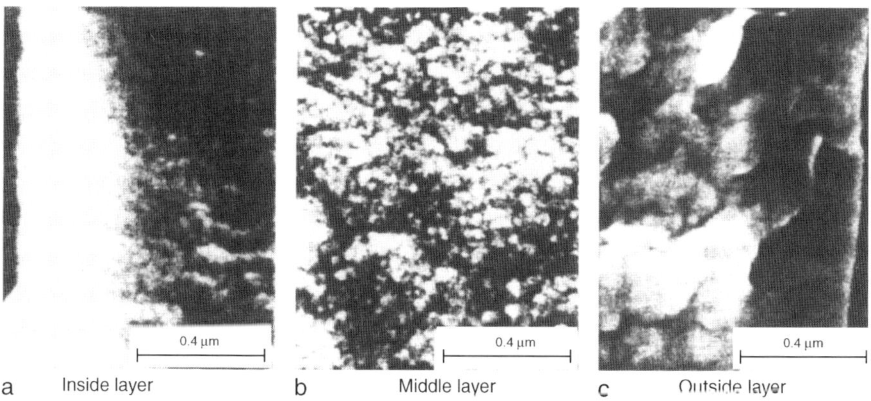

Fig. 6.9. Cross-sectional structure of PMMA B-2 hollow fiber membrane. Reprinted from [13]. Copyright 1992, with kind permission from the Society of Polymer Science, Japan

6.2.2 Cross-sectional Images of Membranes by AFM

As mentioned earlier, the study of the cross-sectional structure of membranes by AFM was hampered by difficulties involved in making a smooth cross-sectional surface. Wood [14] made some earlier studies on the cross section of poly(phenylene oxide) (PPO) hollow fibers. But her attempt did not provide clear information on the nodular structure.

Khulbe et al. [4] conducted an AFM study of the cross section of UF poly(etherimide) hollow fibers, fabricated by the dry–wet spinning method at various air gaps.

Even though it was difficult to produce a smooth surface area of sufficiently large size by slicing the fiber with a sharp edge (razor blade) or fracturing the fiber at liquid nitrogen temperature, Khulbe et al. [4] were successful at obtaining AFM images at less than 3 μm scan ranges. The thicknesses of the walls of hollow fibers were in a range from 76 to 68 μm. The fiber piece of less than 2 mm in length was glued vertically on the AFM disc, and the laser beam was focused at the required position of the cantilever before taking the AFM images.

Figure 6.10 shows the AFM image (scan range 2 μm) of the midsection of the wall of a hollow fiber that was fabricated at a 50-cm air gap. From Fig. 6.10 it seems there are spheres in the cross section, which are designated as nodules due to their dimensions, as suggested by Kesting [15]. Some nodules are fused with each other to form nodule aggregates. In Fig. 6.10 there are dark areas, which may be macrovoids, or paths of pores that are commonly observed by SEM.

Figure 6.11 shows the cross section of the wall of a hollow fiber fabricated at a 30-cm air gap. There are layers of nodules in rows, but in the middle there is a long,

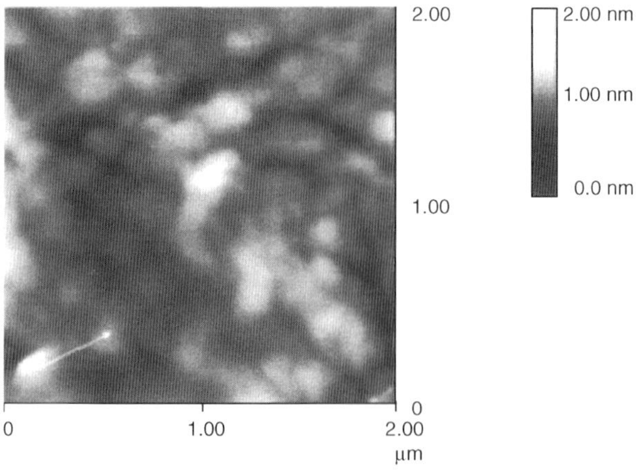

Fig. 6.10. AFM image of the cross section of a UF poly(etherimide) hollow fiber membrane fabricated at a 50-cm air gap. The sample was prepared by cutting the hollow fiber with a sharp edge. The *white arrow* shows the direction from the inner surface toward the outer surface. Reprinted from [4]. Copyright 2006, with kind permission from Elsevier

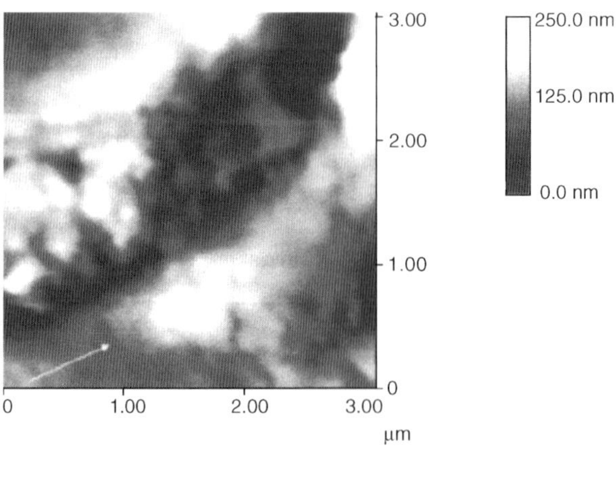

Fig. 6.11. AFM image of the cross section of a UF poly(etherimide) hollow fiber membrane (middle section) fabricated at a 30-cm air gap. The sample was prepared by cutting the hollow fiber with a sharp edge. Reprinted from [4]. Copyright 2006, with kind permission from Elsevier

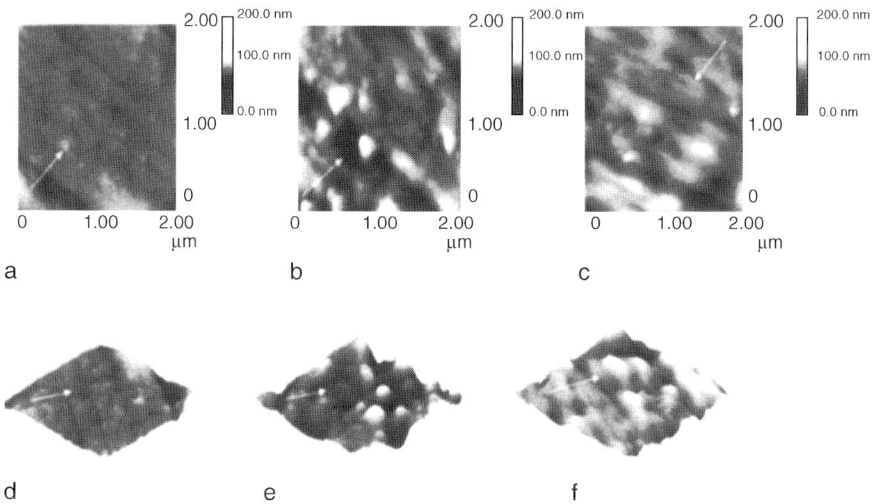

Fig. 6.12a–f. AFM images of the cross section of a UF poly(etherimide) hollow fiber membrane fabricated at a 70-cm air gap **a** near the inner surface, **b** at the middle section, and **c** near the outer surface, and three-dimensional AFM images **d** near the inner surface, **e** at the middle section, and **f** at the outer surface. Samples were prepared by cutting the hollow fiber with a sharp edge. Reprinted from [4]. Copyright 2006, with kind permission from Elsevier

dark area, which could be the area of a finger-like macrovoid. This is commonly observed in hollow fibers. Similar AFM images were also obtained for other hollow fibers prepared at different air gaps.

Figure 6.12a–c shows the AFM images near the inner surface, middle section, and near the outer surface, respectively, of the hollow fiber fabricated at a 70-cm air gap. Figure 6.12d–f shows the three-dimensional AFM images near the inner surface, middle section, and near the outer surface, respectively, of the same hollow fiber. The *arrow* in Fig. 6.12a shows the direction from the inner surface toward the outer surface. Fig. 6.12a suggests that nodules are in a row, in the direction perpendicular to the arrow, and are compacted in comparison with the middle section (Fig. 6.12b) and with the area near the outer surface (Fig. 6.12c). The area near the outer surface seems very coarse, and the nodules are fused with each other. The dark space indicates the pores. Similar AFM pictures were obtained for the other studied membranes. The surprising observation is that the nodules are not at random as reported by Fujii et al. [13], but aligned to the angular direction. The AFM picture of the middle section is very similar to those observed by Fujii et al. [13] in the middle section of Cuprophan, PMMA B-1, and PAN hollow fiber membranes when studied by FE-SEM technique.

Khulbe et al. [4] also measured the nodule sizes (the average of at least 20 measurements) at the inner surface, at the areas near the inner surface, in the middle section and near the outer surface, and at the outer surface for hollow fibers fabricated at different air gaps. Table 6.1 summarizes the results.

Table 6.1. Average nodule size at the inner and outer surfaces and in the cross section near the inner surface, middle section, and near the outer surface of UF poly(etherimide) hollow fibers for different air gaps

Air gap (cm)	Mean nodule size (nm)				
	Inner surface	Cross section			Outer surface
		Near the inner surface	Middle section	Near the outer surface	
10	42.9	30.5	64.3	51.0	43.5
30	36.0	35.8	65.5	40.0	40.0
50	47.8	44.0	64.3	36.8	34.0
70	43.4	34.5	76.5	33.9	33.5
90	–	33.1	87.5	34.0	30.8

From Table 6.1, the following can be concluded:

1. The nodule sizes are similar to those measured by Fujii et al. [13].
2. The nodule sizes at the inner surface and near the inner surface are similar. There is no significant effect of air gap on the nodule sizes observed in this region.
3. The nodules in the middle section are larger than the other regions. The nodule sizes in the middle section increase as the air gap increases.
4. The nodule sizes at the area near the outer surface and at the outer surface are similar. The nodule sizes in this region decrease with an increase in air gap.

The first observation mutually confirms the nodules sizes observed either by SEM or by AFM. The second observation is expected because the coagulation of the polymer occurs soon after the extrusion of the casting dope from the spinneret, when it is brought into contact with the bore fluid (water). The third observation is similar to the result obtained by Fujii et al. [13] for their PAN hollow fiber, in which the mean diameters of the nodules, called spheres in their paper, are reported at the area near the inner surface, in the middle section, and at the area near the outer surface to be 118.0, 166.0, and 109.0 nm, respectively. The fourth observation is probably due to the stretching of the outer surface that occurs while the hollow fiber is traveling through the air gap.

In summary, the following observations were made on the morphology of the PEI hollow fibers fabricated by the dry–wet spinning technique:

1. The cross section near to the inner layer contains well-defined nodules with a smaller number of nodules and nodule aggregates.
2. The maximum nodule size occurs in the middle section.
3. Nodules are aligned in rows in the angular direction in the inner, middle, and outer sections of the hollow fiber cross section.
4. The area near the outer surface is very porous and nodules are fused with each other. Very few isolated nodules are observed.

Several investigators have reported the nodular alignment to the axial direction in hollow fibers [9, 12, 16, 17]. However, the observation of nodular alignment to the angular direction is quite surprising. This suggests the presence of a stress in the angular direction in the spinning dope before the dope is extruded from the spinneret. The reason for the angular stress is unknown. Such a stress is supposed to be absent when the annular space through which the spinning dope is forced to move is perfectly symmetrical. We speculate that the injection point of the bore fluid is slightly off-centered.

There could be other reasons for the angular stress, such as:

1. The fibers did not go into the coagulation bath perpendicularly, but rather in a conical way in a circular motion. This circular motion could be one of the reasons for the angular stress working during the polymer coagulation.
2. Immediately after the casting solution came out of the spinneret, the inner surface of the hollow fiber was brought into contact with an internal coagulant (water), and the solvent/nonsolvent (water) exchange started to take place. In this process, the stream of water entering the hollow fiber wall would form a whirlpool around the nodules, which may have resulted in the stress in the angular direction.

Khulbe et al. [4] also attempted to prepare the AFM sample by fracturing a hollow fiber at liquid nitrogen temperature. Figure 6.13 shows the AFM image of the cross section near the inner surface of the hollow fiber fabricated at a 70-cm air gap. Figure 6.13 also shows clearly that the nodules are aligned in rows in the angular direction. This confirms the cross-sectional structure given in Fig. 6.10, which shows the pictures that were taken from the sample prepared by cutting a hollow fiber with a sharp edge. In Fig. 6.13, it seems that the nodules are slightly stretched (elongated), which could be due to the bending forces applied when the fiber was fractured at the experimental condition (liquid nitrogen temperature).

On studying the cross-sectional area near the inner (active layer) surface, in the middle section, and near the outer surface by fracturing the hollow fiber at liquid nitrogen temperature, the observed results were similar to those obtained by cutting the hollow fiber with a sharp edge. Moreover, cutting with a sharp edge could provide larger smooth areas.

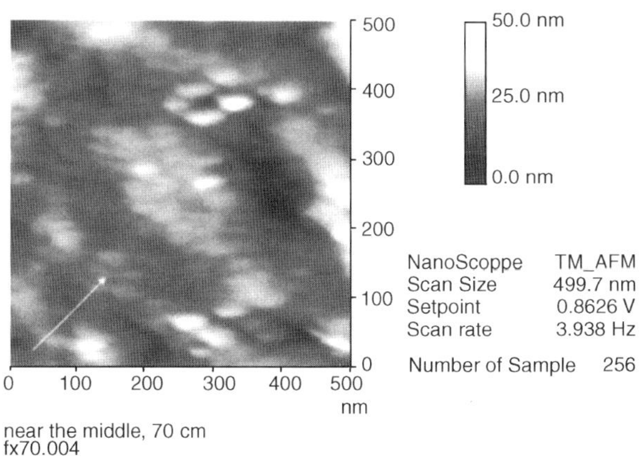

Fig. 6.13. AFM image of the cross section of a UF poly(etherimide) hollow fiber membrane fabricated at a 70-cm air gap. The sample was prepared by fracturing the hollow fiber at liquid nitrogen temperature. The *white arrow* shows the direction from the inner surface toward the outer surface. Reprinted from [4]. Copyright 2006, with kind permission from Elsevier

Figure 6.14 shows the cross section of the thick, dense PPO film (thickness 1.5 mm), which was fabricated by spreading the casting solution containing 15 wt.% PPO in trichloroethylene (TCE) on a glass plate, then evaporating the solvent at room temperature for 5 days. The arrow in the figure shows the direction from the top surface toward the bottom surface. A rectangular piece of the sample approximately 2 × 3 mm was cut with a sharp razor. One side of the longer cross section was glued

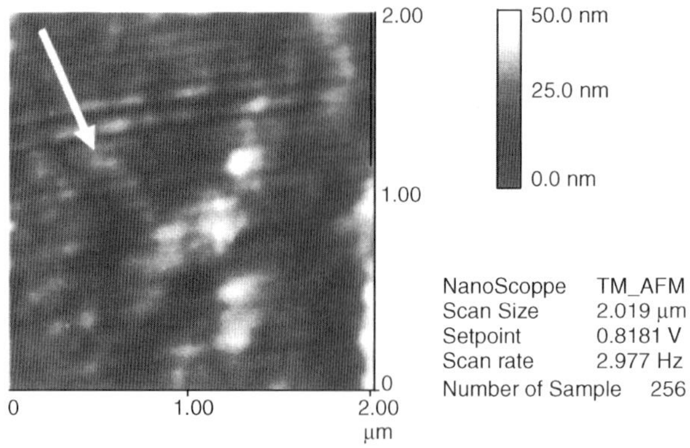

Fig. 6.14. A cross-sectional AFM image of thick, dense PPO film (thickness 1.5 mm). Reprinted from [18]

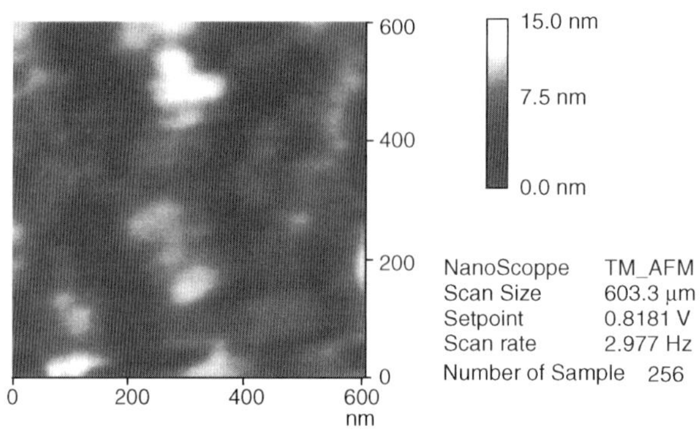

Fig. 6.15. A cross-sectional AFM image near the top surface of PPO film in a smaller scan range [18]

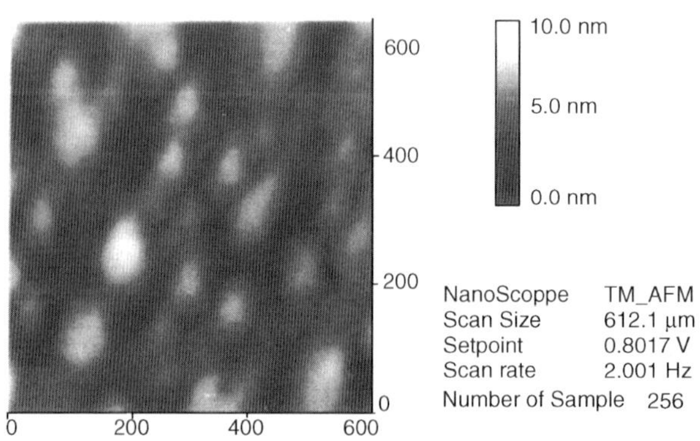

Fig. 6.16. A cross-sectional AFM image in the middle area of PPO film [18]

on the AFM disk, so the other side of the cross section could be studied by AFM. Fig. 6.14 shows that the nodules are in rows and are not randomly distributed [18]. A comparison of the compactness of the nodules reveals such compactness decreases with an increase in distance from the top surface. Figure 6.15 shows the AFM image of the area near the top surface in a small scan range. Figure 6.16 shows the AFM image of the middle range of the cross section. It is clear that the nodules are bigger (Fig. 6.16) in the middle section than the nodules near the skin layer (Fig. 6.15). It should be noted that nodules in the middle section are in layers.

Figure 6.17 shows the AFM image of the cross section near the top layer of the asymmetric poly(ether sulfone) membrane that was as thick as 1.5 mm. The polymer solution was cast on a glass plate at room temperature. The solvent was evaporated at room temperature for a few seconds, and then the film was immersed, together with the glass plate, in ice-cold water for at least 24 h. The film was washed with distilled water and dried at room temperature for three days. The membrane thus prepared had an asymmetric structure. The AFM image shown in Fig. 6.17 is that of the area near the top surface. This figure clearly shows that the nodules are assembled in layers. The arrow in the figure shows the direction from the top surface toward the bottom surface. It seems that compactness of the nodules in the layer decreases toward the bottom, which is expected. From this observation, it seems that during evaporation of

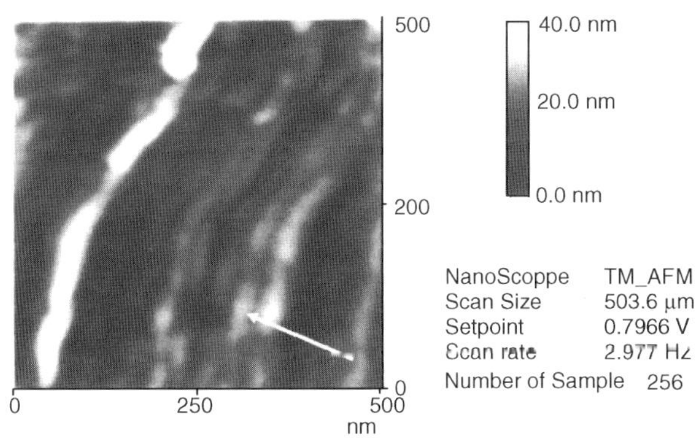

Fig. 6.17. A cross-sectional AFM image near the top surface of an asymmetric poly(ether sulfone) membrane (thickness 1.5 mm) [18]

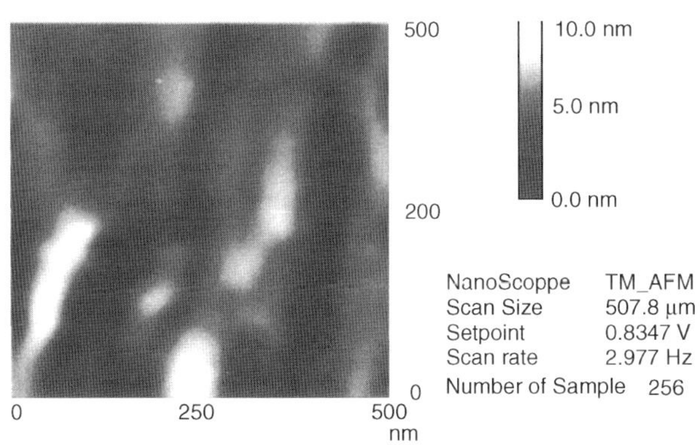

Fig. 6.18. A cross-sectional AFM image near the bottom surface of an asymmetric poly(ether sulfone) membrane (thickness 1.5 mm) [18]

the casting solution, nodules are formed in a regular way, i.e., layer by layer. The dark regions in the picture may be parts of pores. The width of the dark region increases in the depth direction. Figure 6.18 shows the AFM image of the cross section near the bottom of the membrane. In this figure, the nodules are fused together, forming layers as observed near the top skin layer. The whole picture is a scenario of the porous structure, which was expected.

From these studies on flat sheet membranes, the following two general conclusions can be reached:

1. In flat sheet membranes, the nodules/nodule aggregates observed in the cross section are aligned parallel to the surface-forming layers. In hollow fibers, the alignment of the nodules was observed in the angular direction. In hollow fiber formation, the rate of the bore fluid influences the alignment of the nodules.
2. In flat sheet membranes, the nodules are smaller and tightly packed near the top region. The size increases in the depth direction.

6.3 Summary

The studies of membrane morphology by SEM have produced a large number of cross-sectional pictures for polymeric membranes since the onset of asymmetric cellulose acetate membranes by Loeb and Sourirajan. The contribution of those pictures to the design of novel membranes with improved performance was truly phenomenal. SEM requires cumbersome sample preparation, which may hinder true images. AFM does not need such sample preparation, and the pictures taken by AFM are considered to reflect the true nature of membrane morphology.

As discussed, the cross-sectional view of the membranes observed by AFM has similar characteristics to those observed by high-resolution FE-SEM, confirming that AFM can be used to study the cross-sectional structure of polymeric membranes, particularly in terms of their nodular structures. The void spaces between the nodules may form water channels in reverse osmosis and ultrafiltration. They may also become defects when they appear at the densely packed monolayer of nodules or nodular aggregates. Information on the nodular structure will therefore help to eliminate the unwanted defects in the skin layer of the asymmetric membranes.

One of the important features of AFM that cannot be easily utilized in the AFM study of cross-sectional structures is the roughness parameters, since the surface roughness of the sample depends on the method of slicing or fracturing the membrane. As well, nanoscale or submicroscale void spaces have not yet been observed in the cross section. The AFM applications in the study of the cross-sectional morphology have only begun. More efforts in this research field are called for.

References

1. Hachisuka H, Ohara T, Ikeda K (1999) A new type of asymmetric polyimide gas separation membrane having ultrathin skin layer. In: Pinnanau I, Freeman BD (eds) Membrane formation and modification. ACS Symposium Series 744. American Chemical Society, Washington, DC, p 74

2. Park HC, Moon YS, Rhee HW, Won J, Kang YS, Kim UY (1999) Effect of solvent exchange on the morphology of asymmetric membranes. In: Pinnanau I, Freeman BD (eds) Membrane formation and modification. ACS Symposium Series 744. American Chemical Society, Washington, DC, p 121
3. Panar M, Hoehn HH, Herbert RR (1973) Macromolecules 6:777
4. Khulbe KC, Feng C, Matsuura T, Khayet M (2006) Desalination 201:130
5. Kesting RE, Fritzsche AK (1993) Polymeric gas separation membranes. Wiley, New York, pp 264–265
6. Wienk IM, Boomgaard ThVD, Smolders CA (1994) J Appl Polym Sci 53:1011
7. Boom RM, Wienk IM, Boomgaard ThVD, Smolders CA, (1992) J Membr Sci 73:277
8. Kawakami H, Mikawa M, Nagaoka S (1997) J Membr Sci 137:241
9. Pinnau I, Koros WJ (1991) J Appl Polym Sci 43:1491
10. Pinnau I, Koros WJ (1992) J Membr Sci 7:81
11. Xu ZK, Shen LQ, Yang Q, Liu F, Wang SY, Xu YY (2003) J Membr Sci 223:105
12. Fusaoka Y, Imazu E, Fujii Y (1988) Polym Prepr Jpn 37:2579
13. Fujii Y, Iwatani H, Kigoshi S (1992) Polymer J 24:737
14. Wood M (2002) Ph.D. thesis, University of Ottawa
15. Kesting RE (1990) J Appl Polym Sci 41:2739
16. Barzin J, Feng C, Khulbe KC, Matsuura T, Madaeni SS, Mirzadeh H (2004) J Membr Sci 237:77
17. Lee SH, Kim JJ, Kim SS, Kim UY (1993) J Appl Polym Sci 49:539
18. Khulbe KC, Feng CY, Matsuura T, Khayet M (2006) unpublished work

7 Adhesion

7.1 Introduction

Although membranes are widely used for seawater desalination, wastewater treatment, drinking water production, and many other industrial and medical applications, a major obstacle for the efficient application of membrane technology is the phenomenon of membrane fouling. Fouling results in deterioration of membrane performance and ultimately shortens membrane life. Thus, understanding the causes of membrane fouling and developing strategies for fouling control and cleaning are major challenges. Adhesion of particles on the membrane surface is the main cause for fouling. The property of adhesion is also important in membrane science for fabricating composite membranes.

Adhesion is defined as the physical attraction or joining of two substances, especially the macroscopically observable attraction of dissimilar substances. There are many techniques to study adhesion, namely pull-off tests, interfacial fracture tests, blister tests, mapping of interfacial properties, probe modification, and scratch tests.

Van der Waals forces are always present between molecules or between particles and may be attractive or repulsive [1,2], depending on whether they are working between like materials or dislike materials. For like materials, the van der Waals forces are always attractive; however, repulsive forces are predicted for certain dissimilar material combinations [3]. Van der Waals forces have been used to explain why neutral chemically saturated atoms congregate to form liquids and solids. These forces are a main reason for fouling of membranes. Using AFM, repulsive van der Waals forces can be measured with higher precision than attractive van der Waals forces [3,4].

Membrane separation processes are used to separate ions, colloids, and biological molecules from the fluid stream. For optimum operation, the membrane has to possess physical properties, giving appropriate interactions with solutes in the process stream. The most important properties are pore size distribution, surface morphology, appropriate electrical double layer interactions, and minimum adhesion (fouling) [5]. Within the framework of the DLVO theory [6] by the interplay between van der Waals forces and the electrical double layer force, surface interactions can be explained. (DLVO theory is an acronym for a theory of the stability of colloidal dispersions developed by Derjaguin, Landau, Verwey, and Overbeek [6,7]. The theory was developed to predict the stability against aggregation of electrostatically charged particles in dispersion [1].) AFM has been widely used to measure DLVO-type interactions [5,8–12] between a single colloid particle and a normally flat surface, as a function of separation distance. The electrical properties of membrane surfaces have been most commonly evaluated by electrokinetic techniques such as streaming potential measurements [12], but they have some limitations [5]. The adhesion of many polymers is still not clear at a nanometric scale. Surface and interface properties can be modified by a change in chemical composition or structure.

Several devices for measuring surface forces have been developed, including the surface force apparatus [13, 14], the force balance [15], the osmotic stress method, and the total internal reflectance microscope [16]. But in all these methods there are limitations.

Atomic force microscopy is now used for the measurement of adhesion forces. In fact, atomic force microscopic studies can be divided into topographical applications (imaging mode) and force spectroscopy, or so-called atomic force spectroscopy (AFS), i.e., measuring forces as a function of distance [17]. In the former, one generates an image of the sample surface to observe its structural or dynamic features, which has been employed very successfully on a wide variety of surfaces, including polymers [18–20] with resolutions in the micrometer to subnanometer ranges, thus facilitating imaging at the submolecular level.

There are two methods to measure adhesion forces by AFM:

Contact mode AFM In contact mode AFM, the tip is mechanically contacted with the sample surface under a defined applied force. This applied force can be estimated from a force–distance curve, which is obtained by extending the tip to the surface to make contact between the tip and the surface, followed by retracting the tip.

Figure 7.1 shows the force–distance curve. There is no interaction between the tip and the surface when the tip is far away from the surface (*A* in Fig. 7.1). When the tip is close to the surface, there is an attractive force between them. At some point, the force gradient becomes larger than the spring constant of the cantilever, so the tip snaps to the surface (*B*–*C*).

Once the tip is in contact with the surface, cantilever deflection will increase as the end of the cantilever is brought closer to the sample. If the cantilever is sufficiently stiff, the probe tip may indent the surface at this point. In this case, the slope or shape of the contact part of the force curve can provide information about the elasticity of the sample surface. Extending the tip (along line *C*–*D*) results in loading (repulsive) forces to the surface. These repulsive forces are usually used as a feedback parameter for the AFM system to obtain surface morphology.

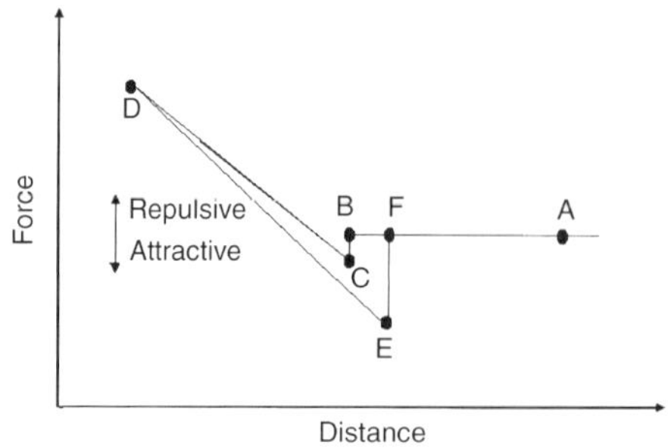

Fig. 7.1. Contact mode AFM force–distance curve

After loading the cantilever to a desired force value, the process is reversed. As the cantilever is withdrawn (along line D–E), adhesion or bonds formed during contact with the surface can cause the cantilever to adhere to the sample some distance past the initial contact point (corresponding to B and C) on the approach curve.

In the retracting cycle, because of the adhesion properties established after the contact between the tip and the surface, the tip will depart from the surface when the force used to pull the tip from the surface exceeds the adhesion force between them (from E–F). This pull force can be considered a measure of the adhesion force between the tip and the surface [21–25]. The adhesion force can be related to the surface energy of the sample surface [23, 24].

Dynamic force mode AFM This mode is developed for the soft features. An oscillating cantilever is used to image surface features in dynamic force mode. There are two main types of dynamic mode [17].

1. In *tapping mode* (TM-AFM), the cantilever is deliberately excited by an electrical oscillator to amplitudes of up to approximately 100 nm so that the cantilever effectively bounces up and down as it travels over the sample. The oscillation amplitude is measured as an *rms* (root mean square) value of the deflection detector signal. The feedback system is set to detect the perturbation on the oscillation amplitude caused by intermittent contact with the surface. In liquids, the tip of the cantilever taps the sample gently during part of the force curve; this mode is similar to the tapping mode operating in air, except that the sample is tapped against the tip instead of the cantilever being driven at resonance to tap the sample.
2. In *non-contact tapping mode* (NC-AFM), the oscillating cantilever does not touch the surface of the sample. The spacing between the tip and the sample is on the order of tens to hundreds of angstroms, with an oscillation amplitude only about 5 nm. In this mode, a sinusoidal excitation of the cantilever with a frequency close to its main resonant is involved. In order to excite the vibration of the probe, in some applications, it is convenient to externally modulate the long-range probe–sample interactions. Therefore, the relatively long-range attractive forces induce changes in the amplitude, frequency, and phase of the cantilever and maintain a constant distance during scanning. These changes in amplitude or in phase can be detected and used by the feedback loop to produce topographic data.

However, there are many disadvantages for the measurement of adhesion properties by AFM:

1. The tip radius must be known.
2. There should be no plastic deformation of the sample surface.
3. Smooth surfaces are required.
4. Due to higher bending modes of the cantilever, AFM can cause misleading results.
5. Small asperities can significantly reduce adhesion.

7.2 Study of Adhesion Forces by AFM

Li and Elimelech [26] wrote an excellent article on the organic fouling and chemical cleaning of nanofiltration membranes (measurements and mechanisms). Leite and Herrmann [17] wrote a fine review on the application of atomic force spectroscopy (AFS) for the study of adhesion phenomenon. In the latter review, they stated that the AFS technique became a useful tool to investigate adhesion forces. AFS forces can help explain how the magnitude of the adhesion force depends on long- and short-range interactions and can offer information on the tip–sample contact area.

Recently, Bowen et al. [27,28] and Hilal and Bowen [29] and Hilal et al. [30] applied the AFM technique to study adhesion at the membrane surface. The measurement of interaction forces between a colloid probe and a membrane surface allows quantification of the electrostatic double layer interactions when the colloid probe approaches the membrane surface, and of the adhesion force (van der Waals interaction force) when the colloid probe is withdrawn after it has been in contact with the membrane surface. Quantification of the interaction forces involved in fouling and chemical cleaning of fouled membranes is very important in order to understand the mechanism of fouling and to develop a favorable membrane for water treatment.

Aimé and his coworkers [31] attempted to show macroscopic models can be successfully applied to measure the adhesion forces by AFM. Butt et al. [16] wrote a review on measuring surface forces in an electrolyte solution with AFM. Local surface properties like the surface charge density or micromechanical properties can be determined by AFM.

For microscopy, the most commonly used probes are the sharp microfabricated silicon nitride or silicon tips. These tips provide high resolution when measuring surface forces. But it is difficult to determine the surface geometry of these probes as the tip area is less than 10 nm [31,32]. Besides this, the surface chemistry of silicon nitride tips, which are most frequently used for force measurement, is very complex [33–35].

The problems of unknown surface geometry and chemistry of the conventional probes can be solved by attaching smooth micrometer-sized spheres to the cantilever. For a sphere of radius R' and a flat surface, a simple equation (Eq. 7.1) relates the measured force F to the interaction energy per unit area V_A [16]:

$$V_A = F(D)/2\pi R' \tag{7.1}$$

where D is the distance between interacting surfaces. From this relationship, it can also be seen that a larger radius results in a larger force F for a given interaction energy V_A. Thus, the system may have a higher sensitivity for the interaction energy. This is the reason why a sphere with a radius of 2–10 μm is attached for adhesion study, while microfabricated tips used for topographic study have a radius of only 5–50 nm. Another advantage of using spheres as probes is the possibility of making a variety of probes of different chemical compositions readily available to the microfabricated cantilevers.

Polymeric films (or membranes) are favorable materials to study adhesive properties with AFM. Electronic controls and software used to record the force-versus-distance curves, revealing the interaction between the surface of the sample and the tip, are now commercially available.

To measure the adhesion forces between colloids (fouling) and the surface layer, or between the surface layer and the substrate, AFM is seen as reliable and simple, as discussed earlier. Bowen et al. [36] introduced a new technique for the direct measurement of the force of adhesion of a single particle by AFM. AFM, in conjunction with the colloid probe technique, allows a direct quantification of membrane fouling through the measurement of pull-off (detachment) forces, when the probe is retracted from the surface after contact has been made.

The school of Bowen [36] used the AFM technique for direct measurement of the adhesion force between a colloidal probe and membrane surfaces. Colloidal probes were prepared by attaching an 11 μm polystyrene sphere with epoxy resin to a V-shaped AFM tipless cantilever (Fig. 7.2). The AFM allowed the measurement of the force between the colloidal probe and a membrane sample as a function of the displacement of the sample, where a piezoelectric crystal varied the sample displacement. A laser beam reflected from the back of the cantilever fell on a split photodiode that detected small changes in the deflection of the cantilever. To convert the deflection to a force, it was necessary to know the spring constant of the cantilever and to define the zero of the force. The spring constant specified by the manufacturer was $0.4\ \mathrm{N\,m^{-1}}$. The zero of the force was chosen where the deflection was independent of the piezo position (where the colloidal probe and the membrane surface were far apart.)

The authors studied two membranes. Both were manufactured by PCI Membrane Systems (UK). The first, ES 404, was made from poly(ether sulfone) and the other, XP 117, was made from a mixture of polymers chosen to achieve low membrane fouling. Both had the nominal MWCO of 4000 Da. The ES 404 membrane had a smaller mean pore size and a narrower pore size distribution (mean pore size 0.65 ± 0.19 nm, maximum pore size 0.95 nm, minimum pore size 0.22 nm) than the XP 117 membrane (mean pore size 1.05 ± 0.30 nm, maximum pore size 1.54 nm, minimum pore size 0.53 nm). Analytical results of the surface morphology of the membranes are given in Table 7.1.

Figures 7.3 and 7.4 show the plots of normalized force (force/particle radius) versus piezo displacement as a colloid probe was retracted from the membrane surface for ES 404 and XP 117, respectively. In making such measurements, the colloid probe first had to be brought into (momentary) contact with the membrane surface. It was reported that the loading forces for such contact have an influence on the measured adhesive force [22,37]. The loading forces for the curves shown are 223.5 nN (Fig. 7.3)

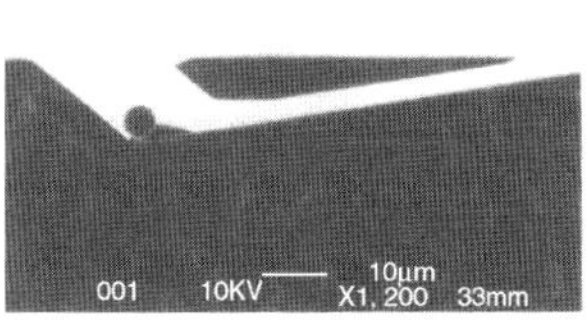

Fig. 7.2. Scanning electron micrograph of a microfabricated silicon nitride cantilever commonly used for force measurements. (A polystyrene microsphere glued at the end of the V-shaped cantilever can be seen pointing upward. The scale bar corresponds to 10 μm.) Reprinted from [10]. Copyright 1992, with kind permission from the American Chemical Society

Table 7.1. Surface morphology analysis of poly(ether sulfone) membranes over an area 20 × 20 nm [36]

Membrane	MWCO	Mean height (nm)	Peak to valley (nm)	Average roughness (nm)	R_q[a] (nm)
ES 404	4000	0.92	1.70	0.15	0.20
XP 117	4000	0.55	1.10	0.09	0.11

[a] Root mean square of Z values

and 221.9 nN (Fig. 7.4), i.e., they were essentially identical. The nature of these plots is best understood by considering Fig. 7.3. From A to A', the colloid probe and membrane move together with no displacement relative to the piezo. From A' to B, there is a stretching of the probe and/or membrane, which gives them a relative movement with respect to the piezo. From B to C, the stretching continues, and the contact between the colloid probe and the membrane is finally broken at C. From C to D, the probe and membrane move farther apart. The difference in force between B and C is a direct measurement of the adhesive interaction; in this case, it is 1.98 mN m^{-1}.

Figure 7.4 shows that the adhesive interaction between the polystyrene colloid probe and the XP 117 membrane under the same conditions is very much weaker. It

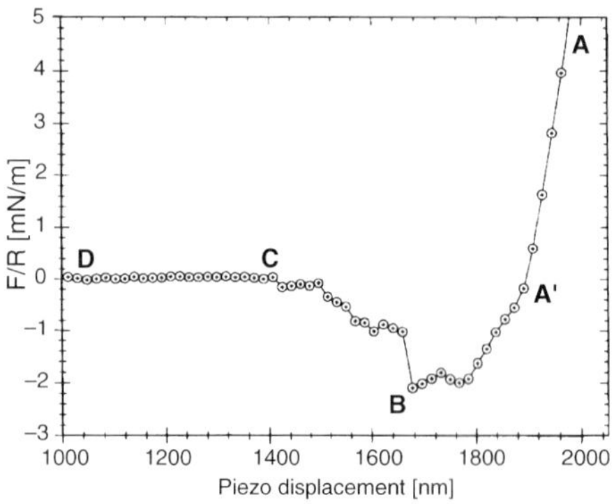

Fig. 7.3. Normalized force-versus-piezo displacement plot (retraction) for a polystyrene colloid probe and an ES 404 membrane (NaCl concentration 10^{-2} M, pH 8.0). Reprinted from [36]. Copyright 1998, with kind permission from Elsevier

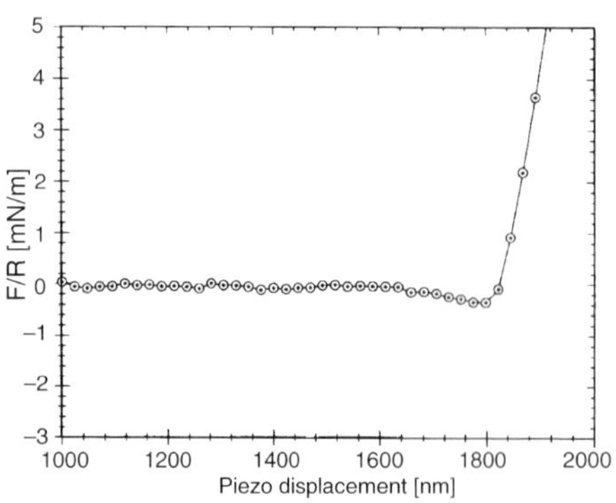

Fig. 7.4. Normalized force-versus-piezo displacement plot (retraction) for a polystyrene colloid probe and an XP 117 membrane (NaCl concentration 10^{-2} M, pH 8.0). Reprinted from [36]. Copyright 1998, with kind permission from Elsevier

is only 0.38 mN m^{-1}, more than five times less than the value for ES 404. As a result, there is also much less stretching of the probe–membrane contact as the piezo is retracted. The results show that the manufacturer has genuinely produced a membrane to which the polystyrene particle weakly attaches. That is, the membrane has a potential for genuinely low-fouling process application. The development of such a technique (sensor: colloid probe) for quantifying the adhesive force at membrane surfaces allows a relatively fast procedure for assessing the potential fouling of membrane surfaces by particles.

Liu et al. [38] studied the interaction forces between bovine serum albumin (BSA) and alkylated chitosan membranes, cross-linked by glucose aldehyde, in media of different pH. The force-displacement curves were obtained by using BSA-coated tips. The results indicated that, at the isoelectric point (IP, pH 4.8), the sample membrane exhibited higher adhesion forces, and deviation from the IP, i.e., at pH 2 and pH 6, decreased the adhesion forces. The adhesion forces at pH 2 were, however, less than those at pH 6 due to the presence of electrostatic repulsive (at pH 2) and attractive (at pH 6) interactions. It was confirmed qualitatively from this study that the introduction of hydrophobic side chains (up to C8) to chitosan could facilitate protein adsorption. However, a long, flexible side chain (C16) can suppress protein adsorption to a certain degree. It was proposed that the protein adsorption could be tuned by adjusting the lengths of the introduced side-alkyl moieties grafted to chitosan: (1) incorporation of alkyl chains with some lengths can promote the protein adsorption and (2) a too-long, flexible side chain tends to hinder protein adsorption.

Bowen et al. [39] measured directly the adhesion (interaction) of cellobiose and cellulose with two polymeric UF membranes of similar MWCO, but of different materials. As probes, they used silica spheres (diameter 5–8 μm) the surfaces of which were modified by static adsorption of cellobiose. They also used pure cellulose probes. Membrane ES 404 was made of poly(ether sulfone) alone, and EM 006 was made of a poly(ether sulfone)-polyacrylate blend, chosen specifically to increase the hydrophilic properties and decrease the fouling properties of the membrane. Study of ES 404 and EM 006 had shown that the interaction of cellobiose or of colloidal cellulose with the membranes was such that ES 404 always had the greater adhesion and greater fouling tendency. However, if the membrane was first fouled with cellobiose, the colloidal cellulose adhesion force was increased significantly, and the differences between the membranes diminished. Bowen et al. suggested that in the future, it would be possible to use the techniques developed to allow prior assessment of the fouling propensity of process streams with different types of membranes.

Mizes et al. [40] used AFM force measurements to perform adhesion studies on doped and undoped polycarbonate surfaces. The doped films were prepared by dissolving N,N'-diphenyl-N'-bis(3-methylphenyl)-(1,1'-biphenyl)-4,4'-diamine (TPD) and polycarbonate at a ratio of 2:3 by weight in a methyl chloride solvent.

The force exerted by the surface on the sharp tip (SiN_4) mounted on the cantilever beam was monitored by AFM. The authors concluded:

1. Adhesion measurement is repeatable.
2. The special variation in adhesion depends on both macroscopic topography and material properties.

The magnitude of adhesion was found to decrease as the local curvature on the surface increased. Their study also suggested that certain topographic features, such as scratches, could lead to increased adhesion values because of the extended area of interaction between the tip and the probed surface.

Eaton et al. [41] studied the surface heterogeneity of a polymer-blend film of polymethyl methacrylate (PMMA) and poly(dodecyl methacrylate) (PDDMA) by adhesion force distribution using AFM (TopoMetrix TMX2000 Discover Scanning Probe Microscope, ThemoMicroscopes, Bicester, UK.). V-shaped cantilevers with a length of 200 μm and a nominal spring constant of 0.032 N m^{-1} bearing an integrated standard profile tip (part no. 1520-00; ThemoMicroscopes, Santa Clara, CA) were used. Films were prepared by dip-coating glass slides into a stirred solution of PMMA and PDDMA in chloroform. From their work, it was concluded that there was a direct relationship between the topographic features on the surface and the relative force of adhesion associated with each domain. For example, pits in the topography image corresponded to high adhesion. It was also demonstrated that the average force of adhesion to PDDMA was much higher than to PMMA.

Hilal et al. [30] used AFM, in conjunction with a colloid probe, coated colloid probe, and cell probe techniques, to measure directly the adhesive force between two different UF membranes and a polystyrene sphere (diameter 11 μm), protein bovine serum albumin (BSA), and a yeast cell. These two membranes were ES 404 and XP 117 mentioned above (Table 7.1). The experiments were performed in 10^{-2} M NaCl solution. It was reported that the adhesive force of the polystyrene, the protein, and the cell system on the ES 404 membrane was greater than that on the XP 117 membrane. The relatively high affinity of protein for synthetic membrane surfaces was also observed.

Nie et al. [23] generated localized shear deformations (scratches) on the surface of a biaxially oriented polypropylene (PP) film using a stylus-type surface profiler, in which a tip with a radius of 2.5 μm was used to scan the film surface. It was observed that the adhesion force increased in the deformed area. This could be due to a buildup of extra free energy, which might be from part of the work done by the mechanical scratching in increasing density and in the ordering of polymer strands on the deformed surface. It was also observed that the physical modification of surface energy by mechanical scratching was different from the increase in surface energy by UV/ozone treatment, which induced chemical functional groups on the surface through surface oxidation. Thus, the energy of a polymer surface can be locally changed by mechanical scratching without changing the surface energy of the surrounding polymer and without influencing the bulk. This could have a number of effects, including creating *local* active sites on surfaces, in terms of surface energy increase or wettability improvement.

Li and Elimelech [26] studied the adhesion forces between the bulk foulant and the fouling layer using AFM on a thin film composite NF membrane (NF-270, FilmTec Corp., Minneapolis, MN), which is a semi-aromatic, piperazine-based PA membrane). A carboxylate-modified colloid probe was used as a surrogate for humic acid, the major organic foulant in natural waters. A correlation was observed between the measured adhesion forces and the fouling and cleaning behavior of the membrane under various solution chemistries. The AFM measurements confirmed that the Ca^{2+} ions greatly enhanced natural organic (NOM) fouling by complexation and subsequent formation of intermolecular bridges among organic foulant molecules.

Nishimura et al. [9] measured interacting forces between a polymethyl methacrylate sphere and a flat plate, onto which expandable fluorine mica platelets (EM, Co-op Chemical Co. Ltd., Tokyo) were coated. By using the AFM technique, Nishimura et al. [9] successfully measured the forces interacting in a NaCl aqueous solution between the PMMA probe and the EM-coated flat surfaces that apparently behaved as a macroscopic basal plane of EM. The data for forces obtained by the above technique were consistent with the forces theoretically calculated according to the DLVO theory.

Gordano et. al. [42] studied three HYFLON AD [copolymers of tetrafluoroethylene (TFE) and 2,2,4-trifluoro-5-trifluoromethoxy-1,3-dioxole (TTD)] composite membranes under NaCl solution (0.1 M) by AFM. These composite membranes were made of HYFLON AD 60X, supported on polyamide microfiltration membranes with a pore size of 0.2 μm from AKZO. A dilute solution of HYFLON was knife-cast on a polyamide support membrane, and the solvent was evaporated. Membranes dried at −5, 4, and 25 °C were named X, Y, and Z, respectively. Colloid probes were used by attaching a polystyrene sphere with a typical diameter of 10.5 μm to the end of a V-shaped AFM tipless cantilever. Similarly, silica spheres with a typical diameter of 6.0 μm were attached. Adhesion of polystyrene and silica particles to the membranes was measured in NaCl solutions at three different concentrations to quantify their likely fouling tendencies. To measure the adhesion force, the colloid was first brought into momentary contact with the membrane surface with a specified loading force, 12–140 nN, for all measurements. The loading force may have some influence on the magnitude of the adhesive force [22]. Normalized adhesion forces (force/sphere radius, F/R) for polystyrene and for silica for all membranes under all conditions are given in Table 7.2. The Z membranes had the lowest overall force, Y had intermediate, and X the greatest adhesion. The adhesion increased with an increase in NaCl concentration from 10^{-4} to 10^{-3} M, except for the silica probe on the X membrane. All membranes showed greater adhesive interactions with the polystyrene sphere than have been reported for the polysulfone membrane, which gave an adhesive interaction of 1.98 mN m^{-1} in 10^{-2} M NaCl [36]. There are several possibilities to explain this difference. First, the polysulfone membrane was smoother (0.7 nm) than the composite membranes (3.1–15.2 nm). Second, the different materials have different surface chemistries and hence different interactions. Overall, lower adhesion was measured for silica than for polystyrene.

Table 7.2. Normalized adhesion force for a polystyrene and silica colloid probe at the membrane [42]

Membrane	Electrolyte concentration (M)	Mean F/R (force/sphere radius) ($nN\,m^{-1}$) Polystyrene	Silica
Z	10^{-4}	5.3 (± 0.8)	0.3 (± 0.2)
Z	10^{-3}	8.4 (± 0.7)	1.7 (± 0.8)
Z	10^{-2}	8.2 (± 2.0)	2.2 (± 0.8)
Y	10^{-4}	3.8 (± 0.5)	2.0 (± 0.5)
Y	10^{-3}	3.5 (± 0.6)	3.0 (± 0.7)
Y	10^{-2}	4.0 (± 0.6)	2.8 (± 0.9)
X	10^{-4}	8.0 (± 4.1)	8.2 (± 2.5)
X	10^{-3}	8.7 (± 2.4)	6.2 (± 3.6)
X	10^{-2}	17.6 (± 5.9)	4.2 (± 1.4)

The work of Gordano et al. [42] suggests that a number of factors might influence the adhesion of particles to membranes, including electrostatic interactions, van der Waals forces, short-range interactions, surface roughness, and membrane deformability. For quantitative interpretation of data, the influence of each of the important factors must be considered.

Weisenhorn et al. [43] studied adhesion, attraction, and repulsion between surfaces in liquids with an atomic force microscope. Their work also shows the usefulness of the AFM for investigations of micromechanical properties.

7.3 Summary

Atomic force microscopy is an important tool and widely used to take images of the membrane surfaces in the nanoscale range. It also allows characterization of the adhesive behavior on a nanoscale level. In particular, the colloid probe technique in AFM is a powerful method for the characterization of interfacial phenomena by measuring forces and distances between two surfaces. Thus, fouling of membrane surfaces by colloidal particles can be predicted from the results of AFM experiments. The knowledge acquired by this method will also help the development of new membrane materials with low or zero fouling properties. Furthermore, quantification of colloid–membrane interaction forces is useful for the development of theoretical predictions and, thus, optimization and control of membrane separation processes. So far, no research has been done on the adhesion of the top layer with the substrate in composite membranes. However, the study of adhesion by AFM for composite membranes would help to understand adhesion between the top dense and the porous substrate membranes. Only rough estimates of the interfacial energy can be given by AFM due to the limitation of AFM techniques presently available.

References

1. Israelachvili JN (1997) Intermolecular and surface forces. Academic, London
2. Hunter RJ (1989) Foundations of colloid science.Oxford University Press, New York
3. Lee S, Sigmund WM (2001) J Colloid Interface Sci 243:365
4. Lee S, Sigmund WM (2003) Colloids Surf A 204:43
5. Bowen WR, Doneva TA, Austin J, Stoton G (2002) Colloids Surf A 201:73
6. Verwey EJW, Overbeek JThG (1948) Theory of the stability of lyophobic colloids. Elsevier, Amsterdam
7. Derjaguin BV, Landau L (1941) Acta Physicochim URSS 14:633
8. Bowen WR, Hilal N, Lovitt RW, Wright CJ (1999) Colloids Surf A 157:117
9. Nishimura S, Kodama M, Noma H, Inoue K, Tateyama H (1998) Colloids Surf A 143:1
10. Ducker W, Senden T, Pashley R (1992) Langmuir 8:1831
11. Larson I, Drummond C, Chan D, Grieser F (1997) Langmuir 13:21
12. Nyström M, Lindström M, Matthiasson E (1989) Colloids Surf A 36:297
13. Tabor D, Winterton FRS, Winterton RHS (1969) Proc R Soc London Ser A 312:435
14. Israelachvili JN, Adams GE (1978) J Chem Soc Faraday Trans 74:975
15. Derjaguin BV, Rabinobich YI, Churaev NV (1978) Nature 272:313
16. Butt HJ, Jaschke M, Ducker W (1995) Bioelectrochem Bioenerg 38:191
17. Leite FL, Herrmann PSP (2005) J Adhes Sci Technol 19:365
18. Souza NC, Silva JR, Pereira-da-Silva MA, Roposo M, Faria RM, Giacometti JA, Oliveira ON Jr (2004) J Nanosci Nanotechnol 4:1
19. Job AE, Herrmann PSP, Vaz DO, Mattoso LHC (2001) J Appl Polym Sci 79:1220
20. Riul A, Dhanabalan A, Cotta MA, Herrmann PSP, Mattoso LHC, MacDiarmid AG, Oliveira ON (1999) Synth Met 101:830
21. Radmacher M, Fritz M, Cleveland JP, Walters DA, Hansma PK (1994) Langmuir 10:3809
22. Toikka G, Hayes RA, Ralston J (1996) J Colloid Interface Sci 180:329
23. Nie HY, Walzak MJ, Berno B, McIntyre NS (1999) Appl Surf Sci 144–145:627
24. Nie HY, Walzak MJ, Berno B, McIntyre NS (1999) Langmuir 15:6484
25. van der Vegte EW, Hadzioannou G (1997) Langmuir 13:4357
26. Li Q, Elimelech M (2004) Environ Sci Technol 38:4683
27. Bowen WR, Hilal N, Lovitt RW, Sharif AO, Williams PM (1997) J Membr Sci 126:77
28. Bowen WR, Hilal N, Jain M, Lovitt RW, Sharif AO, Wright CJ (1999) Chem Eng Sci 54:369
29. Hilal N, Bowen WR (2002) Desalination 150:2889
30. Hilal N, Bowen WR, Lovitt RW, Wright C (2002) Eng Life Sci 2:131
31. Aimé JP, Elkaakour Z, Odin C, Bouhacina T, Michel D, Curély J, Dautant A (1994) J Appl Phys 76:754
32. Siedle P, Butt HJ, Bamberg E, Wang DN, Kuhlbrandt W, Zach J, Haider M (1992) Inst Phys Conf Ser 130:361
33. Drummond CJ, Senden TJ (1994) Colloids Surf A 87:217
34. Harame DL, Bouse LJ, Shott JD, Meindl JD (1987) IEEE Trans Electron Devices 34:1700
35. Sende TJ, Drummond CJ (1994) Colloids Surf A 94:29
36. Bowen WR, Hilal N, Lovitt RW, Wright CJ (1998) J Membr Sci 139:269
37. Veeramasuneni S, Yalmanchili MR, Miller JD (1996) J Colloid Interface Sci 184:594
38. Liu WG, Li F, Zhao XD, Yao KD, Liu QG (2002) Polym Int 51:1459
39. Bowen WR, Stoton JAG, Doneva TA (2002) Surf Interface Anal 33:7
40. Mizes HA, Loh KG, Miller RJD, Ahuza SK, Grabowski EF(1991) Appl Phys Lett 59:2901
41. Eaton PJ, Graham P, Smith JR, Smart JD, Nevell TG, Tsibouklis J (2000) Langmuir 16:7887
42. Gordano A, Arcella V, Drioli E (2004) Desalination 163:127
43. Weisenhorn AL, Maivald P, Butt HJ, Hansma PK (1992) Phys Rev B Condens Matter Mater Phys 45:11226

8 Membrane Surface Morphology and Membrane Performance

8.1 Introduction

The use of AFM to study surfaces provides information about pore size distribution, surface morphology and electrical properties, surface adhesion/membrane fouling behavior, and the correlation between membrane characteristics and process behavior. This information, in conjunction with mathematical models and performance data, allows for the development of a novel approach in the prediction of new desired membranes.

We discussed in earlier chapters that nodules, nodule aggregates, pore sizes, pore size distributions, and roughness parameters can all be observed on the membrane surface by AFM and quantified using software. Thus, AFM has proved to be a very powerful tool to study membrane surface morphology. However, this does not necessarily satisfy the knowledge-seeking scientist. Although the above parameters are highly valuable for the purpose of membrane characterization, they are of little use, at least for separation membrane scientists and engineers, unless correlations can be found between those parameters and membrane performance. Information on membrane surface characterization will be complete only when the cause-and-effect relationships among membrane preparation, membrane morphology, and membrane performance are fully understood.

Therefore, an attempt will be made in this chapter to find some relationships between the surface characterization parameters obtained by AFM and the membrane performance data. Most obviously, the pore size and the pore size distribution will have a direct influence on the selectivity and the permeation rate of NF, UF, and MF membranes, where pores are most visible.

Furthermore, some interesting discussions can be found in the literature, although still controversial, on the relationship between the roughness parameter and the flux, with respect to RO and gas separation membranes, where the measurement of the pore size is not at all easy. The roughness parameters are also related to the fouling tendency of the RO/NF and UF membranes. It has to be emphasized that all these discussions are found in the papers published during the past decade since the emergence of AFM as a surface characterization tool. Some of the discussions have already provided important guidelines for the sophisticated design of separation membranes. Further progress is expected to take place during the next decade.

8.2 Relationship Between Membrane Morphology and Membrane Performance

8.2.1 Reverse Osmosis and Nanofiltration Membranes

Hirose et al. [1] suggested an approximately linear relationship between membrane surface roughness and permeate flux for TFC RO membranes with six different cross-linked aromatic polyamide skin layers. On a polysulfone (PSf) substrate, cross-linked aromatic polyamides were formed at a water/solvent interface using *m*-phenylene diamine (MPD) and 1,3,5-benzenetricarbonyl trichloride as monomers. Isopropyl alcohol content in the aqueous amine phase was changed from 0 to 60 wt.% to control the interfacial surface tension, which eventually led to different surface roughness values.

The results of their work are given in Table 8.1. A linear relationship was found between the flux and the surface roughness, which was attributed to enlargement of the effective membrane area.

Lu et al. [2] fabricated TFC NF membranes, in which the skin layer was either polyesters or polyamides. The monomers used in the polycondensation reactions are as follows: (1) alcohol (bisphenol-A, BPA), (2) amine (metaphenylene diamine and piperazine), and (3) acid chloride (isophthaloyl chloride, terephthaloyl chloride [3,4], and trimesoyl chloride). They reported that the composite layer (active layer) was smoother than that of the substrate membrane. Upon formation of the active layer, the pore size decreased, which resulted in a flux decrease and a retention increase. Although it is unclear, they seem to maintain that their results confirm Hirose et al.'s conclusion, i.e., the flux increases with an increase in surface roughness [1].

Another kind of TFC NF membrane was studied by Hamza et al. [5] using AFM. Membranes were prepared by applying a thin coat of sulfonated poly(phenylene oxide) solution to a porous substrate poly(ether sulfone), followed by solvent evaporation. Mixtures of chloroform/methanol with different ratios were used as solvents. The authors reported that the nodule size decreased with an increase in chloroform concentration in the solvent mixture. In the separation experiment of sodium chloride solute, the flux decreased from 11 to less than $2 \times 10^{-6}\ m^3\ m^{-2} s^{-1}$ as the chloroform concentration increased from 0 to 66%. Thus, the decrease in flux parallels the decrease in the nodule size. Although they did not report the surface roughness,

Table 8.1. Separation performance of RO membranes and roughness parameters

Sample	Isopropyl alcohol (wt.%)	Rejection (%)	Flux ($m^3\ m^{-2}\ day^{-1}$)	R_a (nm)
RO-1	0	99.8	1.0	43.5
RO-2	5	99.8	1.2	52.5
RO-3	10	99.8	1.3	62.6
RO-4	20	99.7	1.7	87.1
RO-5	40	92.1	1.2	59.8
RO-6	60	53.4	0.52	16.9

in general, the roughness parameter decreases with the decrease in the nodule size. Thus, their data indirectly supports Hirose et al.'s concept [1].

Stamatialis et al. [6] studied cellulose-based dense and asymmetric membranes used for NF and RO. The surface structure of dense and integrally skinned cellulose acetate (CA) and cellulose acetate butyrate (CAB) membranes, prepared by the phase inversion technique under different casting conditions, was investigated by TM-AFM. They stated that for CA membranes, increasing evaporation times and organic pore formers led to higher apparent rejection and lower fluxes. These were also correlated to lower values of the roughness parameters R_a and R_z. The CAB membranes always display much higher apparent rejection than the CA membranes prepared under the same casting conditions. Such higher apparent rejections are correlated to lower R_z values. Their conclusion seems to support Hirose et al.'s results, although the higher rejection of CAB is obviously due to the effect of cellulose chemistry.

Kwak and Ihm [7] used AFM and solid state NMR spectroscopy to characterize structure–property–performance correlations in high-flux RO membranes. The membranes were thin film composites , whose thin active layers were based on aromatic polyamide formed by the interfacial polymerization of MPD and trimesoyl chloride (TMC). These membranes, each coded as SH-I, SH-II, and SH-III, were provided by Saechan (Yongin-city, Korea). The variations among these commercial membranes are difficult to know. Most likely, they vary by the amount of catalyst or surfactant added to the aqueous MPD solution. Table 8.2 shows water flux, salt rejection, and the roughness parameter of those membranes, together with the data for another membrane, MPD/TMC, which was prepared at the laboratory of Kwak and Ihm [7].

From Table 8.2, it seems difficult to correlate the water flux with the surface roughness or the surface area. The MPD/TMC membrane has a higher surface area but lower water flux than SH-I and SH-II, although their surface roughness is almost identical. The difference in the surface roughness and area between SH-I and SH-II is not large enough to explain the increase of water flux in SH-II relative to SH-I. Kwak and Ihm [7] concluded that the microscopic surface structure is not the only major factor responsible for the RO performance, and a more fundamental understanding of the nature of thin film polymers was required. However, combined results of AFM surface analysis and field emission electron microscopy surface observations revealed that, as was the case in SH-III, both the higher surface roughness

Table 8.2. Water flux, salt rejection, surface area, and roughness parameters of RO membranes

Thin film composite membranes	Reverse osmosis characteristics Water flux (gfd)	Salt rejection (%)	Surface roughness, R_a (nm)	Surface area (μm^2)
MPD/TMC	28	> 96.0	42	171.2
SH-I	28.4	> 99.1	40	147.4
SH-II	37.2	> 98.7	43	148.8
SH-III	45.4	> 98.4	84	180.9

and surface area of the membrane were favorable to the higher water flux. The proton spin lattice relaxation in the rotating frame for the aromatic polyamides in their wet state (i.e., saturated with D_2O) was shown to be sensitive to the water flux and played a significant role in enhancing the membrane permeability, regardless of the surface features.

In another work, Kwak et al. [8] prepared TFC membranes with skin layers formed by interfacial polymerization using unsubstituted and methyl- and halogen-substituted bisphenols and 1,3,5-benzenetricarbonyl trichloride as monomers. They demonstrated that substitution of bisphenol biphenyl rings, with either methyl or halogen, strongly influenced rejection and permeability of aromatic polyester RO membranes. Higher flux and lower rejection were associated with the smoother membrane surfaces obtained from methyl substitution. On the other hand, lower flux and higher rejection were associated with the rougher membrane surfaces resulting from halogen substitution. Thus, Kwak et al.'s work does not always support the result obtained by Hirose et al [1].

Mohammad et al. [9] prepared NF membranes through interfacial polymerization and measured the pore diameters by AFM. However, they did not measure the roughness of the membrane surfaces. In general, it was observed that as the pore size decreased, the water permeability, NaCl rejection, and vitamin B_{12} all decreased. A high charge density NF membrane with a large pore size, for example, can be expected to have higher rejection with a higher permeate flux and, thus, have a higher productivity compared to a tighter membrane with a lower charge density [10].

8.2.2 Ultrafiltration Membranes

Chung et al. [11] spun hollow fiber UF membranes from poly(ether sulfone) and reported the effect of the shear rate, working from the spinneret wall to the spinning dope, on the outer surface morphology of the hollow fibers, which was observed by AFM. They also conducted UF experiments with poly(vinylpyrrolidone) and bovine serum albumin as solutes. Table 8.3 correlates mean roughness with pure water flux and solute separation of the membranes. From the data given in Table 8.3, it is clear and that the flux of water increases and solute separation decreases with an increase in surface roughness. Espinoza-Gömez and Lin [12] reported similar results for UF membranes from acrylonitrile copolymers (discussed in Chap. 5).

Khayet et al. [13] prepared flat sheet UF membranes from poly(etherimide) and correlated their UF performance with the roughness parameter and the pore size obtained by AFM. The preparation of those membranes and their mean pore sizes and roughness parameters were discussed in Chaps. 4 and 5. Table 8.4 shows the results of UF experiments together with the mean pore size and the mean roughness parameter of the membranes. From Table 8.4, it is clear that the pure water flux increases and the solute separation decreases as the pore size and the roughness parameter increase. This conclusion corroborates that of Chung et al. [11]. Similar results were obtained for asymmetric poly(etherimide)UF membranes, the surfaces of which were fluorinated by surface-modifying macromolecules [13].

Table 8.3. Mean roughness, R_a, pure water flux, and solute separation of hollow fiber UF membranes

Fiber [a]	R_a (nm)	Normalized flux [b] ($L h^{-1} m^{-2} bar^{-1}$)	Solute separation (%) [c]			
			PVP 10k [d]	PVP 29k	PVP 40k	BSA 67k [e]
1	2.54	300	8.0	19	23	88
2	1.86	140	38	72	80	97
3	1.52	120	55	87	90	99
4	1.35	97	70	92	94	99
5	1.21	74	76	94	96	100

[a] Shear rate and other properties are mentioned in Table 4.8
[b] Pure water flux was measured at a transmembrane pressure of 1.0 bar
[c] Separation tests were carried out with a solution of 200 ppm
[d] PVP 10k: poly(vinylpyrrolidone), M_w 10 000 Da
[e] BSA 67k: bovine serum albumin, M_w 67 000 Da

Table 8.4. Results of UF experiments: pore size (mean diameter) and roughness parameters of flat sheet PEI membranes

Membrane	PWP ($10^{-5} m^3 m^{-2} s^{-1}$)	Solute separation (%) PEG (35 000 Da)	PEO (400 000 Da)	Pore size, mean diameter (nm)	Roughness mean, R_a (nm)
10U	8.0	31.3	89.7	24.1	66.0
12U	6.9	46.5	90.7	20.8	46.1
15U	6.2	56.1	93.5	16.7	43.6
17U	5.6	65.9	96.7	13.1	36.5
20U	5.1	59.7	–	11.5	37.9

Feng et al. [14] reported the effect of surface morphology of UF poly(etherimide) hollow fiber membranes made by the dry–wet spinning method. In making these membranes, the bore fluid flow rate was changed while other parameters were kept constant. It was observed that as the bore fluid flow rate was increased from 0.1 to 0.4 mL min^{-1}, the roughness of the surface increased, as shown in Table 8.5. The data given in Table 8.5 indicates that the pure water permeation flux increases while the solute separation decreases with an increase in the pore size and the surface roughness. This conclusion is the same as that obtained by Chung et al. [11] and Khayet et al. [13]. It is also supported by the work of Khulbe et al. [15], where PEI hollow fiber UF membranes were spun at different air gaps.

Zhang et al. [16] prepared cellulose membranes by casting cellulose solutions in N-methylmorpholine-N-oxide (NMMO), followed by immersion in aqueous NMMO solutions for gelation. The surface morphology of the membranes was studied by AFM. It was found that increasing the temperature and concentration of the coagulation bath led to larger pore sizes, wider pore size distributions, and higher values of the roughness parameter. The membranes displayed higher fluxes and lower rejection of BSA as the pore size and roughness parameter increased. Thus, there was a clear correlation between the roughness parameter and the membrane flux. The ef-

Table 8.5. Results of UF experiments with poly(etherimide) hollow fiber membranes (operating pressure 30 psi)

Bore fluid flow rate ($mL\,min^{-1}$)	Mean surface roughness (nm)	PWP ($L\,m^{-2}\,h^{-1}$)	PEO separation (%) M_w 100 000 Da	M_w 200 000 Da	M_w 300 000 Da	Pore size, diameter, by transport data (nm)
0.1	1.548	41	25.36	63.78	87.65	22
0.2	1.977	127	17.56	38.67	63.52	29
0.3	2.7192	176	7.65	33.65	57.46	32
0.4	2.723	198	0	18.35	46.83	35

fect of the temperature of the coagulation bath on the membrane morphology was greater than the effect of the NMMO concentration in the coagulation bath.

Hilal et al. [17] studied the surface structure of a poly(ether sulfone) membrane, the surface of which was modified by depositing a molecularly imprinted polymer layer. They measured the pore size and surface roughness by AFM. It was reported that an increase in the degree of modification, given by the weight of the imprinted polymer layer, led to a systematic decrease in pore size and an increase in surface roughness. As the pore size decreased, the relative water flux (the ratio of the water flux of the surface-modified membrane to that of the unmodified membrane) also decreased. In other words, the flux decreased as the surface roughness increased.

8.2.3 Pervaporation membranes

Teng et al. [18] studied pervaporation of water/ethanol mixtures through AAm-p-aramide membranes (aromatic polyamide membranes on which acrylamide was grafted by plasma polymerization). Membranes with different degrees of grafting were prepared, and their surfaces were studied by AFM. The roughness parameters increased with an increase in the degree of grafting. The authors further attempted to correlate the roughness parameter with the performance of the membranes. Figure 8.1 shows the relationship between the surface roughness and the permeation rate as well as the separation factor for pervaporation of 90 wt.% aqueous ethanol solutions through the AAm-p-aramide membranes. From Fig. 8.1, it is clear that both the permeation rate and the separation factor are linearly related to the surface roughness.

8.2.4 Gas separation membranes

Khulbe et al. [19] were probably the first to study the relationship between the surface roughness and the gas permeability of dense homogeneous membranes prepared from poly(phenylene oxide). PPO was dissolved in six solvents of different boiling points and cast into a thin film. The solvent was then completely removed under vacuum. Studying the surface roughness by AFM, they found a nearly straight-line relationship between the roughness parameter and the boiling point (Fig. 8.2). The exceptionally low roughness data corresponds to the trichloroethylene solvent. It was also found that the gas permeability generally increased and the selectivity decreased with an increase in the boiling point (Figs. 8.3 and 8.4). From these data, it can be concluded that permeability increases while the selectivity decreases with an increase in the surface roughness.

Attention should be focused on the extremely high selectivity data of the PPO membrane that was cast from the PPO solution in carbon disulfide (PPO-CS_2). On the surface of this membrane it was observed that several nodular aggregates merged

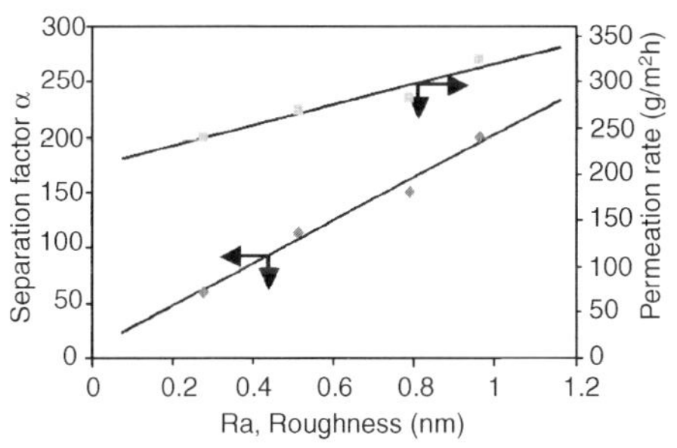

Fig. 8.1. The relationship between the surface roughness and the separation factor and permeation rate for an AAm-p-aramide pervaporation membrane

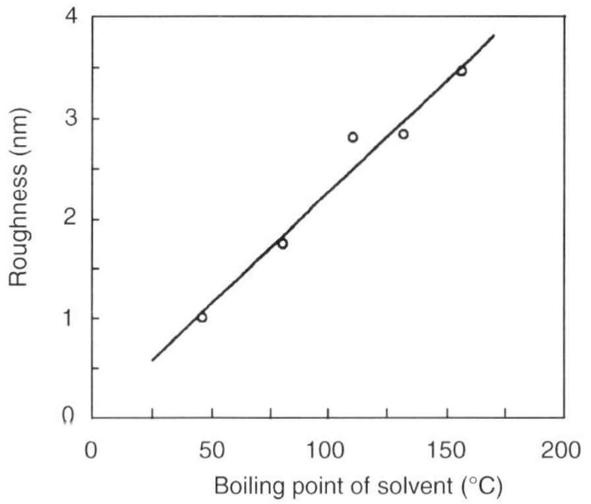

Fig. 8.2. Mean roughness of a poly(phenylene oxide) membrane versus the boiling point of the solvent used for membrane preparation. Reprinted from [19]. Copyright 1997, with kind permission from Elsevier

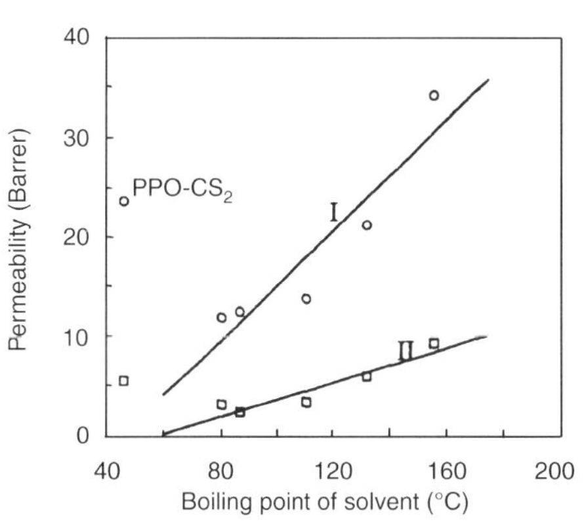

Fig. 8.3. Permeability of O_2 (line *I*) and N_2 (line *II*) versus the boiling point of the solvent used for PPO membrane preparation. Reprinted from [19]. Copyright 1997, with kind permission from Elsevier

to form a large supernodular aggregate. For the other membranes, merging of nodular aggregates did not take place. The exceptionally high selectivity and permeability of the PPO-CS_2 membrane shown in Figs. 8.3 and 8.4 are most likely due to the merging of the nodular aggregates, by which the interstitial void spaces between the nodular aggregates were filled. Combined with plasma etching, it was also shown by Khulbe and Matsuura [20] that the distance between the supernodular aggregates as well as between the nodular aggregates increased in the depth direction of the membrane (Fig. 8.5).

Tan and Matsuura [21] prepared asymmetric poly(phenylene oxide) membranes by the phase inversion technique using chloroform as a solvent. The effect of thirteen additives, branched- and straight-chain alcohols ranging from C_3 to C_{10}, on the separation of O_2/N_2 and CO_2/CH_4 gas pairs was studied. The membrane surface morphology was also investigated by AFM. They found that the roughness parameter of the membrane, R_a, was less than 0.63 nm when the nodules, observed on the membrane surface, were merged, and more than 0.63 nm when they were discrete. The merging of the nodules has an effect on the separation of the above-mentioned gas pairs, as shown in Figs. 8.6 and 8.7. The membranes indicated by *m* (merge) have larger permeance ratios than those indicated by *d* (discrete). This is because the interstitial void spaces between the nodules are filled when the nodules are merged. These are additional experimental data, which show that the smoother membrane enhances the selectivity of the gas mixtures.

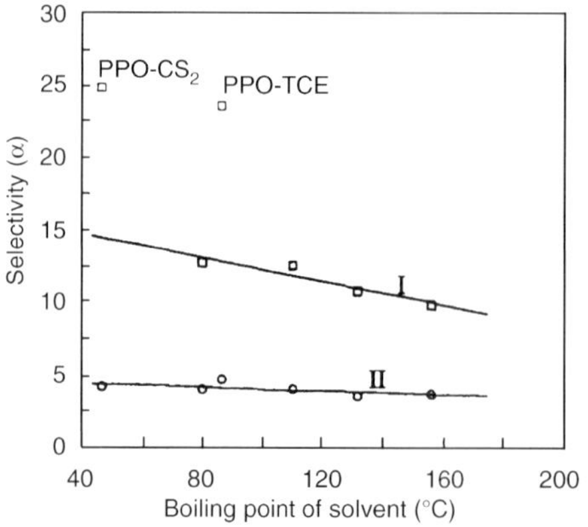

Fig. 8.4. Selectivity of CO_2/CH_4 (line *I*) and O_2/N_2 (line *II*) versus the boiling point of the solvent used for PPO membrane preparation. Reprinted from [19]. Copyright 1997, with kind permission from Elsevier

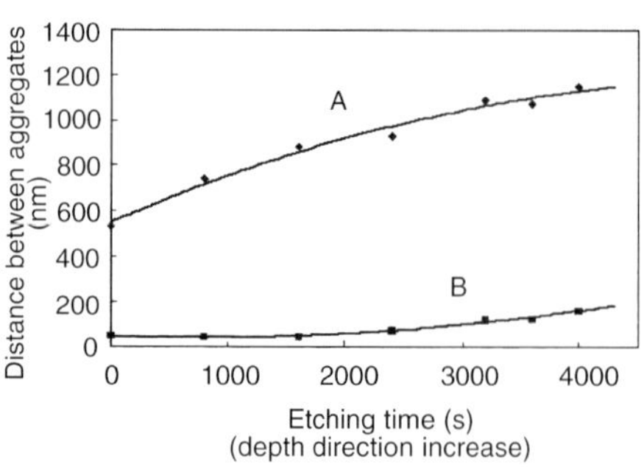

Fig. 8.5. Mean distance between supernodular aggregates at 5-μm scan range (*A*) and between nodular aggregates at 1-μm scan range (*B*) in the PPO-CS_2 membrane after various plasma etching times

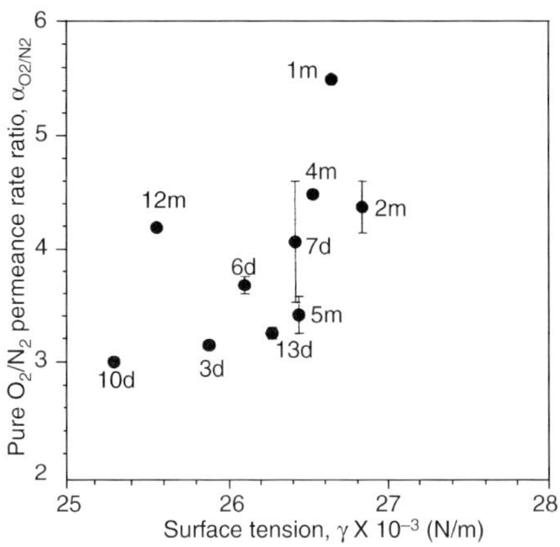

Fig. 8.6. Pure O_2/N_2 permeance ratio of asymmetric poly(phenylene oxide) membranes as a function of surface tension of chloroform/nonsolvent additives mixtures. Nonsolvent additives include 2-ethyl-1-hexanol (1*m*), 1-octanol (2*m*), 2-propanol (3*d*), 2-decanol (4*m*), 3,5,5-trimethyl-1-hexanol (5*m*), 2,4-dimethyl-3-pentanol (6*d*), 2,4,4-trimethyl-1-pentanol (7*d*), 2-methyl-3-hexanol (10*d*), 3-ethyl-3-pentanol (12*m*), and 2-methyl-2-hexanol (13*d*). Merged is indicated by *m*; discrete is indicated by *d*. Reprinted from [22], with kind permission from J. Tan

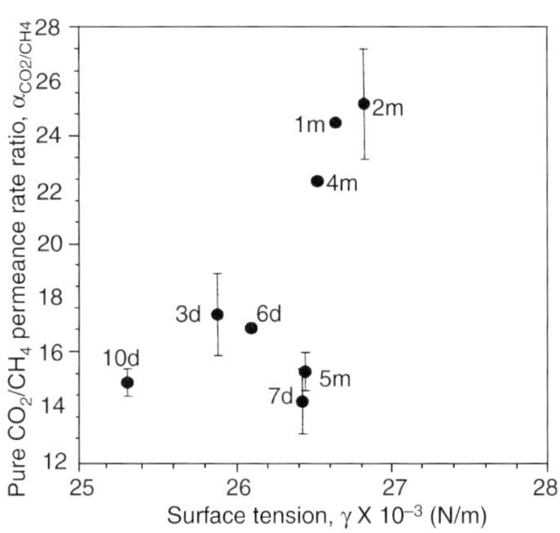

Fig. 8.7. Pure CO_2/CH_4 permeance ratio of asymmetric poly(phenylene oxide) membranes as a function of surface tension of chloroform/alcohol mixtures. Nonsolvent additives include 2-ethyl-1-hexanol (1*m*), 1-octanol (2*m*), 2-propanol (3*d*), 2-decanol (4*m*), 3,5,5-trimethyl-1-hexanol (5*m*), 2,4-dimethyl-3-pentanol (6*d*), 2,4,4-trimethyl-1-pentanol (7*d*), and 2-methyl-3-hexanol (10*d*). Merged is indicated by *m*; discrete is indicated by *d*. Reprinted from [22], with kind permission from J. Tan

Tan [22] also applied plasma etching to the surface of PPO membranes for different time periods and studied gas separation as well as surface roughness. Figure 8.8 shows the roughness parameter, R_a, versus etching time. Figure 8.9 shows that the O_2/N_2 permeance ratio increased initially, followed by a steep decrease as the etching time increased. The initial increase in the selectivity seems to be due to the cross-linking caused by the plasma. In the latter stage of the plasma etching, the selectivity decreased with an increase in surface roughness.

Asymmetric blend poly(ether sulfone)–polyimide (PES-PI) hollow fiber membranes were prepared at different air gaps and used for gas separation [23, 24]. It was observed that the permeance of CO_2 and N_2 increased with an increase in air gap while the ideal selectivity decreased when the hollow fibers were uncoated. After silicone coating, no significant change was observed either in permeance or in the permeance ratio. Although the feed gas was supplied to the shell side of the hollow fiber, it seems that the skin layer is on the lumen side when the SEM pictures are examined. From AFM [25], it was observed that the mean roughness parameter of the inner surface (uncoated lumen side) increased with an increase in the air gap,

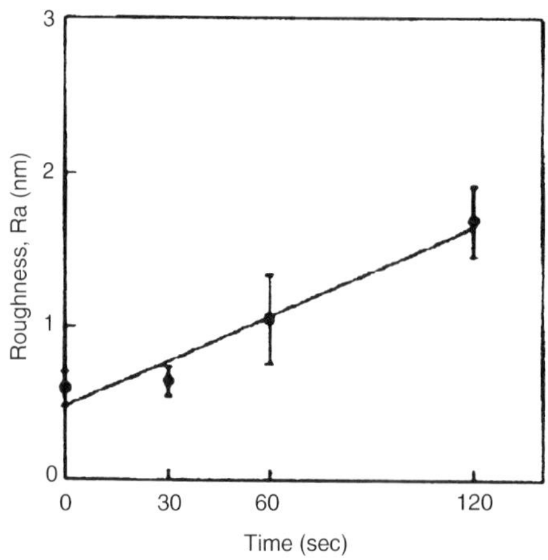

Fig. 8.8. Mean roughness, R_a, as a function of etching time. Reprinted from [22], with kind permission from J. Tan

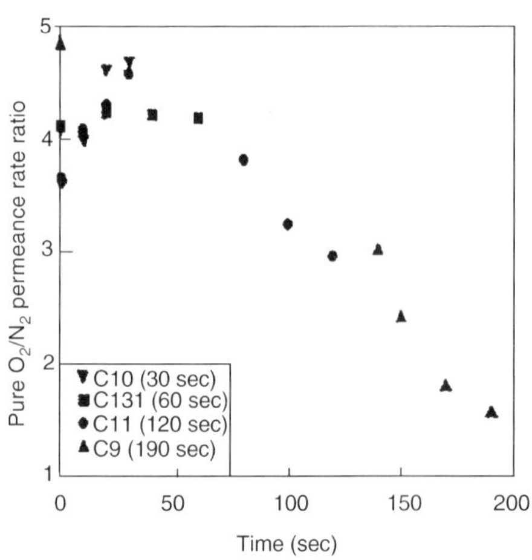

Fig. 8.9. Pure O_2/N_2 permeance ratio versus etching time. Reprinted from [22], with kind permission from J. Tan

while the trend was opposite for the outer surface. If gas permeation is controlled by the inner skin layer, the above observation indicates that the selectivity of the hollow fiber decreased with an increase in the surface roughness parameter. Thus, all the above experimental data support the concept that the selectivity of the gas separation membrane decreases with an increase in membrane surface roughness.

Reid et al. [26] studied the transport properties of surface-treated poly[3-(2-acetoxyethyl)thiophene] (P3AcET) composite membranes. Base- or acid-catalyzed hydrolysis of the ester group of P3AcET yielded poly[3-(2-hydroxyethyl)thiophene], a highly permselective conducting polymer at the membrane surface. Membranes were annealed under vacuum. The surface morphology of P3AcET membranes before and after the base or acid treatment was studied by TM-AFM (Fig. 8.10). TM-AFM of the as-cast membrane (without treatment) revealed a smooth surface as shown in Fig. 8.10a. Base-catalyzed hydrolysis of the membrane produced a dramatic change in surface morphology (Fig. 8.10b), while the measured surface roughness in the AFM image did not increase significantly. The surface microstructure of the acid-hydrolyzed membrane more closely resembled that of the as-cast membrane (Fig. 8.10c). However, the roughness of the surface was the highest among all surfaces. Surface morphology of these membranes was also studied by SEM, and similar results were obtained. Table 8.6 shows the permeance of the as-cast membrane and the membranes with base and acid treatment.

Table 8.7 shows the O_2/N_2 and CO_2/CH_4 permeance ratio for the as-cast P3AcET membrane and the base- or acid-treated membrane. As the table shows, the base or acid treatment resulted in a dramatic increase in the O_2/N_2 permeance ratio, while

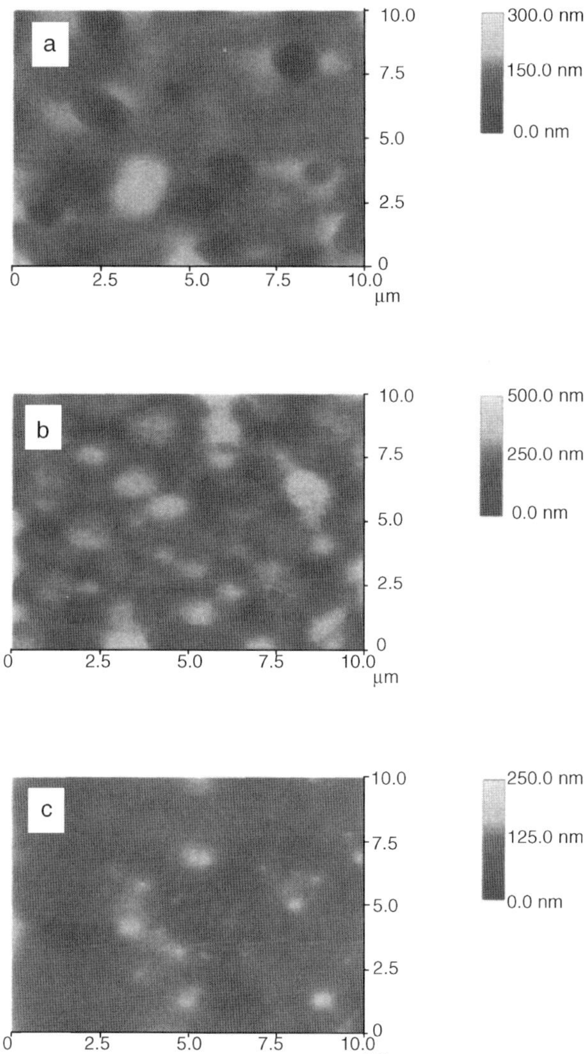

Fig. 8.10a–c. TM-AFM image of P3AcET membrane surface. **a** As cast. **b** KOH modified. **c** H_2SO_4 modified. Reprinted from [26]. Copyright 2002, with kind permission from Elsevier

Table 8.6. Permeance of various gases for the as-cast P3AcET membrane and membranes with base and acid treatment in barrer/cm

Membrane	CO_2	O_2	N_2	CH_4
As-cast	668 ± 40	111 ± 8	22 ± 1	36 ± 3
KOH-treated	281 ± 11	63 ± 7	5 ± 1	14 ± 0
H_2SO_4-treated	336 ± 32	60 ± 6	5 ± 1	8 ± 3

Table 8.7. Permeance ratio for selected gas pairs and roughness parameters

Membrane	R_q (nm)	O_2/N_2	CO_2/CH_4
As-cast	51	5.1 ± 0.5	18.5 ± 1.8
KOH-treated	57	12.9 ± 2.3	20.0 ± 0.8
H_2SO_4-treated	67	11.7 ± 3.8	45.0 ± 13.7

the permeance decreased. Care must be taken in concluding that this is the result of an increase in the roughness parameter, since the material at the membrane surface was changed by the base and acid treatment.

8.2.5 Membranes for Other Membrane Processes

Gould et al. [27] studied the surface of poly(ethylene terephthalate) (PET) films by AFM and reported that the surface roughness could distinguish between the amorphous and crystalline regions. They also suggested that irregularities on the surface could affect the physicochemical properties of the film.

James et al. [28] studied the hydration of Nafion® membranes by AFM and X-ray scattering and reported that tapping mode phase imaging was successfully used to identify hydrophobic and hydrophilic regions of Nafion perfluorosulfonate cation exchange membranes. Since there is often little correlation between the topography and phase images, it is a useful tool for identifying and mapping regions with different properties, irrespective of their topographical natures. The AFM images support the maximum entropy (MaxEnt) [29, 30] interpretation of a cluster model of ionic aggregation, with spacing between individual clusters in the 3–5 nm range, aggregating to form cluster agglomerates in the 5–30 nm range. They also showed that in phase images, the number of clusters decreased and the average cluster size increased with increasing humidity. This supports the interpretation of the MaxEnt charge distributions and is reminiscent of the redistribution of ionic material between clusters proposed in the cluster network model [29, 30].

Soresi et al.'s [31] work shows a different picture. Their work is based on the polystyrene grafting of some PVDF-based membranes followed by sulfonation of aromatic rings. The proton exchange membranes so produced were used for fuel cell research. Figure 8.11 shows the AFM images of the Millipore PVDF membrane (porous) and the laboratory-cast PVDF-HFP copolymer membrane (dense) before and after the grafting reaction. Looking at the surface morphology of the porous

PVDF membrane before grafting (a), the membrane shows a globular structure, in which pores of dimensions around 100 nm are homogeneously distributed. The polymer domains, ranging from 500 nm to 1.0 μm, have a parallel orientation, which is related to the film processing. After grafting, the topography of the matrix changes remarkably (c). The parallel orientation of the domain is compromised by a disordered rearrangement induced by the styrene units grafting onto the PVDF chains. The polymer globule dimensions consequently increase up to 3 μm, as does the average surface roughness. Similar effects on the surface topography due to the grafting process can be detected in the case of the dense copolymer membrane images (b and d). However, the authors did not report the value of the roughness parameter or the diameter of the pores.

Wang et al. [32] sulfonated bisphenol based wholly on aromatic poly(arylene ether sulfone) (PBPS) at various degrees of sulfonation. The surfaces of the membranes prepared from these polymers were studied by AFM. Figure 8.12 shows AFM

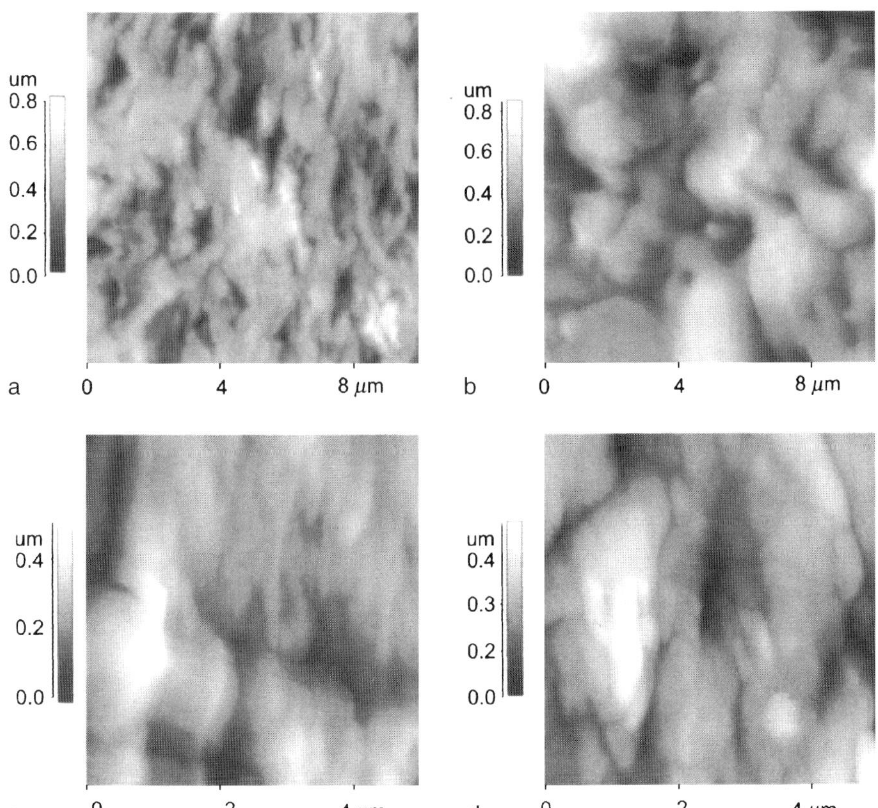

Fig. 8.11a–d. AFM images of a PVDF porous membrane (**a** and **c**, pore dimension 100 nm) and those of a dense PVDF-HFP membrane (**b** and **d**) before grafting (**a** and **b**) and after grafting (**c** and **d**). Reprinted from [31]. Copyright 2004, with kind permission from Elsevier

images of these membranes. For unsubstituted control homopolymers, i.e., PBPS-00, featureless phase morphology was observed. On the other hand, for sulfonated PBPS (designated as PBPSH-*X*, 20–60, where H indicates that the sulfonate is in acidic form and *X* is the degree of sulfonation), dark, cluster-like structures with diameters of 10 – 25 nm are clearly visible in the phase image. The dark area/structures were assigned to a softer region, which represents the hydrophilic sulfonic acid groups containing small amounts of water. The domain sizes and connectivity vary depending on the degree of sulfonation.

For PBPSH-20, an isolated ionic cluster region is formed with a diameter of 10 – 15 nm. In PBPSH-40, the phase contrast of the hydrophilic ionic domains increased and became more easily distinguished from the nonionic matrix, but the domains were still segregated with approximately 25 nm diameters. On the other hand, for the PBPSH-60 sample, the phase image undergoes a significant change, wherein the hydrophilic ionic domains become continuous to form large channels of an ionic-rich phase. Similar continuous ionic channel structures are also observed

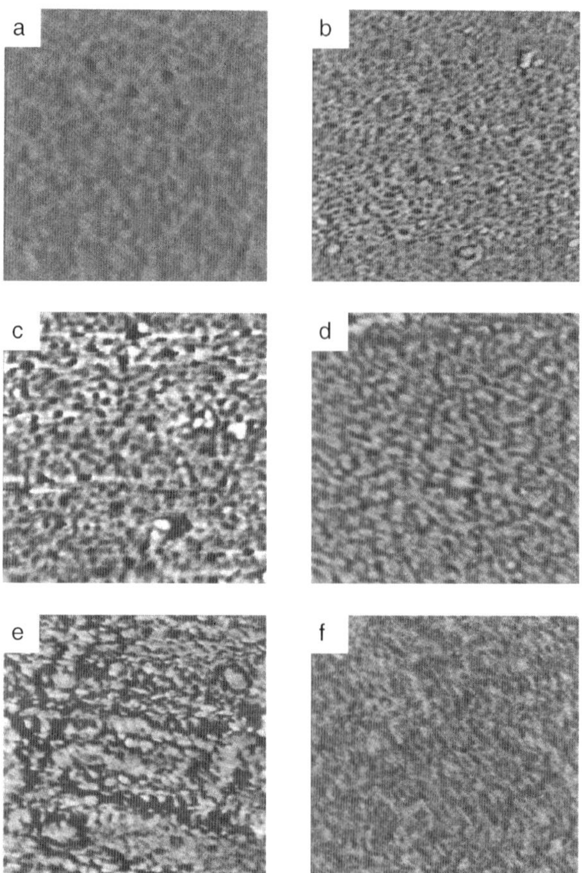

Fig. 8.12a–f. AFM tapping phase images for sulfonated poly(arylene ether sulfone) copolymer and Nafion 117: **a** PBPS-00, **b** PBPSH-20, **c** PBPSH-40, **d** PBPSH-50, **e** PBPSH-60, and **f** Nafion 117 (acid form). Scan boxes are 700 × 700 nm and phase scales are 0–30° for (**a–d**). Scan boxes are 350 × 350 nm and phase scales are 0–10° for (**f**). Reprinted from [32]. Copyright 2002, with kind permission from Elsevier

in the ambient humidity-conditioned Nafion 117 ionomer, although the domain size of the Nafion is about 4–10 nm. Thus, the AFM phase images show that the diameter of the hydrophilic ionic domains increased from 10 to about 25 nm and became connected to produce a co-continuous morphology as the degree of sulfonation increased. This percolation limit of the PBPSH system is found to be about 50 mol.% of the disulfonated monomer. The results from the glass transition temperatures (T_g) of these studied membranes were consistent with the AFM image analysis [16].

Borges et al. [33] studied the tensile properties of cross-linked and un-cross-linked composite films (thickness ~ 20–35 μm) prepared from hydroxypropylcellulose (HPC) with the incorporation of microcrystalline cellulose fibers. The surface morphology of the un-cross-linked and of the cross-linked films with different degrees of fiber incorporation was investigated. Studies in polarizing optical microscopy and AFM seemed to indicate that the tensile properties of these composites were correlated to the packing of fibers. For the concentration of the utilized cross-linking agent, and for the fiber content of 10 wt.%, an optimal packing of fibers throughout the matrix was correlated to the minimal difference between the roughness parameters obtained by AFM analysis of the top and bottom surfaces of the films.

Barzin et al. [34] characterized UF poly(ether sulfone) hemodialysis membranes (Chap. 5). The morphologies of both inner and outer surfaces changed on heating either in hot water or in air, and so did the performances of the membranes. The performance data of hollow fibers heated in air at 150 °C was found to be the most appropriate for hemodialysis application.

8.3 Surface Roughness and Membrane Fouling

Membrane fouling is the primary impediment to the wide use of RO and UF in water treatment processes. Colloidal particles are the major culprits for the fouling of RO membranes although there are other factors, such as sparingly soluble salts, dissolved organic solvents, microorganisms, etc. [35]. During colloidal fouling of RO membranes, colloidal particles accumulate at the membrane surface and increase the resistance to water flow through the membrane. Colloids are ubiquitous in natural and process water. Examples include clays, colloidal silica, oxyhydroxide, large organic macromolecules, organic colloids, suspended matters, and precipitated calcium carbonate. The fundamental mechanism controlling the fouling of the RO membranes is complex and not well understood. To understand the mechanism of fouling and to control it, AFM would be very useful. A number of papers have been published in which the surface roughness and the fouling given as flux decline were correlated.

Perhaps the first paper concerning surface roughness and fouling was that of Elimelech et al. [36], in which they studied the surface morphology of cellulose acetate and composite aromatic polyamide RO membranes by AFM and correlated their findings with membrane colloidal fouling. They observed higher fouling rates for the thin film composite membrane compared to those for the cellulose acetate membrane. The higher fouling rate for the thin film composite membrane was attributed to its greater value of surface roughness, which was inherent in interfacially polymerized aromatic polyamide composite membranes.

This concept was used by some other groups to reduce membrane fouling. For example, Freger et al. [37] studied thin film composite polyamide membranes modified by the graft polymerization of acrylic acid (AA). AFM showed that attachment of the grafted layer onto the active surface did not lead to significant changes in the surface morphology. While a certain reduction of roughness could be obtained with RO membranes, the characteristic "hills and valleys" morphological pattern did not undergo a qualitative change, even for high degrees of grafting. Hence, grafting did not prevent the colloidal fouling of the membrane, whereas it was effective for the reduction of organic fouling.

Vrijenhoek et al. [38] studied the surface morphology, permeability, rejection, and colloidal fouling behavior with respect to two commercial RO and two commercial NF membranes. The RO membranes were Hydranautics LFC-1 (Oceanside, CA) and Trisep X-20 (Goleta, CA). The NF membranes were Dow-FilmTec NF-70 (Minneapolis, MN) and Osmonics HL (Minnetonka, MN). Table 8.8 shows the performance properties and the AFM parameters of the membranes. It is obvious from the data that the authors made RO and fouling experiments at different pressures for different membranes to adjust the initial flux to the same value.

Table 8.8 indicates that there is no correlation between the roughness parameter and the RO data, including pure water permeability and solute separation. On the other hand, colloidal fouling of RO and NF membranes is nearly perfectly correlated with membrane surface roughness, regardless of physical and chemical operating conditions. It was further demonstrated that AFM images of fouled membranes yielded valuable insights into the mechanisms governing colloidal fouling. At the initial stages of fouling, AFM images clearly show that more particles are deposited on the rough membranes than on the smooth membranes. Particles preferentially accumulate in the *valleys* of rough membranes, resulting in *valley clogging*, which causes more severe flux decline than in smooth membranes.

Table 8.8. Performance properties and AFM parameters of membranes

Membrane name (type)	Flux [a] decline J/J_0 (%)	Salt (NaCl) rejection [b] (%)	Pure water permeability (10^{-12} m Pas^{-1})	Average roughness, R_a (nm)
HL (NF)	0.00	35.0	30.7	10.1
X-20 (RO)	5.32	98.0	8.8	33.4
NF-70 (NF)	7.97	83.0	31.9	43.3
LFC-1 (RO)	9.32	98.0	11.0	52.0
Correlation coefficient	1.00	0.87	−0.41	–

[a] Fouling test conditions: $J_0 = 1.415 \times 10^{-5}$ m s^{-1} (30 gfd), ionic strength of electrolyte $I = 0.01$ M NaCl, concentration of silica particles $C_p = 200$ mg L^{-1}, surface velocity $u_{xf} = 0.192$ m s^{-1}, and pH = 6.8

[b] $J_0 = 1.415 \times 10^{-5}$ m s^{-1} (30 gfd), $I = 0.01$ M NaCl, $u_{xf} = 0.192$ m s^{-1}, and pH = 6.8

Table 8.9. Correlation of flux decline data and average membrane surface roughness for systematically varied physical and chemical operating conditions

Membrane name (type)	Flux decline J/J_0 (%) [a]	Flux decline J/J_0 (%) [b]	Flux decline J/J_0 (%) [c]	Flux decline J/J_0 (%) [d]	Average roughness, R_a (nm)
HL (NF)	0.0	13.4	13.9	15.1	10.1
X-20 (RO)	5.32	21.8	38.3	19.4	33.4
NF-70 (NF)	7.97	27.2	46.9	21.4	43.3
LFC-1 (RO)	9.32	28.9	49.3	28.4	52.0
Correlation coefficient	1.00	0.99	0.99	0.93	–

[a] $J_0 = 1.415 \times 10^{-5}$ m s^{-1} (30 gfd), $C_p = 200$ mg L^{-1}, $I = 0.01$ M NaCl, $u_{xf} = 0.192$ m s^{-1}, and pH = 6.8
[b] $J_0 = 1.415 \times 10^{-5}$ m s^{-1} (30 gfd), $C_p = 200$ mg L^{-1}, $I = 0.01$ M NaCl, $u_{xf} = 0.096$ m s^{-1}, and pH = 6.8
[c] $J_0 = 1.415 \times 10^{-5}$ m s^{-1} (30 gfd), $C_p = 200$ mg L^{-1}, $I = 0.05$ M NaCl, $u_{xf} = 0.096$ m s^{-1}, and pH = 6.8
[d] $J_0 = 9.433 \times 10^{-6}$ m s^{-1} (20 gfd), $C_p = 100$ mg L^{-1}, $I = 0.01$ M NaCl, $u_{xf} = 0.125$ m s^{-1}, and pH = 6.8

Table 8.9 shows the correlation of flux decline data and the average membrane surface roughness for systematically varied physical and chemical operating conditions. The data from Table 8.9 prove that the influence of membrane surface roughness is not limited to a specific set of physical and chemical conditions. The first fouling data set is perfectly correlated with membrane surface roughness. The second and third columns of fouling data are nearly perfectly correlated with membrane surface roughness. However, the fourth data set is very highly correlated with membrane surface roughness. In the fourth set, fouling was studied under totally different physical and chemical operating conditions. The flux decline (fouling) data from Table 8.9 suggests that the fouling behavior of these membranes could be qualitatively predicted from membrane surface roughness measurements regardless of the physical or chemical test conditions used.

Zhang et al. [39] studied composite NF membranes prepared by interfacial polymerization of piperazine and trimesoyl chloride and coated with poly(vinyl alcohol) (PVA). AFM showed that the coating was effective in membrane surface smoothing. Even though no data was shown, they argue that the increase in hydrophilicity and smoothness of the membrane surface may lead to reduction of surface fouling based on the previous findings [8].

Bowen et al. [40] used sulfonated poly(ether ether ketone) (SPEEK) as an additive in the polysulfone (PSf)/SPEEK/*N*-methyl-2-pyrrolidone (NMP) system. Membrane characterization was carried out using filtration studies and AFM. Membranes prepared from solutions in the region of *polyelectrolyte behavior* [41] showed more pronounced and systematic improvement of membrane permeability and salt rejection. A small decrease in pore size and surface roughness was also followed by an increase in SPEEK content. Compared with a -28.5 mN m^{-1} adhesion force of a 4 μm silica particle for a SPEEK free PSf membrane, a SPEEK modified membrane showed greatly reduced adhesion of -0.75 mN m^{-1}. This, together with the surface smoothening effect, leads to the reduction of membrane fouling when the surface is modified by the addition of SPEEK.

In another study, Bowen et al. [42] prepared membranes from polymer blends of polysulfone and sulfonated poly(ether ether ketone) (PSf/SPEEK). It was reported that these membranes had high porosities, high charge densities, and pore sizes at the boundary between NF and UF. For comparison, two commercial membranes of cellulose acetate and poly(ether sulfone) were chosen. Therefore, the following four membranes were involved in their study:

1. CA202, a cellulose acetate membrane of nominal MWCO (2000 Da)
2. ES404, an aromatic poly(ether sulfone) membrane of nominal MWCO (4000 Da)
3. S5-20, a flat sheet membrane made from PSf/SPEEK/NMP [43] with a polymer content of 20 wt.%
4. T5-20, similar to S5-20, but with a polymer content in the casting solution of 25 wt.%

Filtration experiments were performed using model water to simulate typical Scottish water composition: 10 mg L^{-1} of humic acid in a 0.001 M NaCl solution. Fe^{3+}, Al^{3+}, and Mn^{2+} ions, in their appropriate forms as chloride salts, were used at concentrations of 1.0, 0.1, and 0.1 mg L^{-1}, respectively.

The results showed that the fouling (defined as flux after 4 hours of filtration followed by rinsing/initial flux) of SPEEK-blended PES membranes was as low as that of the CA membrane, particularly at the lower operating pressure. However, the fluxes of SPEEK-blended PES membranes were much higher than the CA membrane. The fouling of the PES membrane was the worst. The easy rinsing of the SPEEK-blended PES membranes was confirmed by AFM study, since the surface topology before the membrane surface was easily restored by rinsing for the blended membrane, while in the case of the PES membrane, the humic acid deposit stayed on the membrane surface even after rinsing.

In other studies, AFM was used to investigate the structure of the fouling layer. By examining the top surfaces of microfiltration membranes using AFM, before and after fouling by filtration of apple juice, it was found by Riedl et al. [44] that surface roughness was a more influential factor in flux decline than the surface hydrophobicity of the membranes. Poly(ether sulfone) and poly(vinylidene fluoride) (PVDF) membranes showed superior performance, possibly due to their rough surface structure. They suggested that rough membranes produce a looser surface-fouling layer than the dense fouling layers observed on smooth-surfaced membranes such as nylon and polysulfone. Thus, according to Riedl et al., a rough membrane surface is better than a smooth membrane surface [44].

Huisman et al. [45] reported the effect of protein–protein and protein–membrane interactions of membrane fouling in the ultrafiltration of bovine serum albumin. The membranes used were polysulfone UF membranes of different molecular weight cutoffs from Millipore. They showed protein–membrane interactions influence the membrane fouling only in the initial stage. Protein–protein interactions become more important in the latter stage. AFM images of the membrane surfaces taken after the filtration experiments showed that the membranes were totally covered by a fouling layer. The structure of this fouling layer depended strongly on pH. In particular, very open structures with high permeabilities were found at a low pH (below the isoelectric point of the protein). These induced high values of flux and protein transmission.

James et al. [46] studied fouling during filtration of milk by two Osmonics™ flat sheet membranes: an Elements DS-J series MF membrane (pore size 3 μm) and an Elements G-20 UF membrane (molecular weight cutoff 3500 Da). Combined with the SEM and AFM study, they showed that the surface morphology and internal microstructure of a membrane have a great effect on fouling and filtration performance. Initial fouling during filtration of skim milk proceeds by protein–polymer and protein–protein interactions. It has been shown that a gel layer forms on the surface of membrane. This layer is slightly compressible and densifies as it grows. The steady-state permeate flux that is achieved is due to the gel layer permeability. The microstructure of the gel layer itself has been found to influence permeate flux.

Murata and Tanioka [47] studied the interfacial state change of a cellulose triacetate (CTA) membrane by AFM, when negatively or positively charged polyelectrolytes were adsorbed onto the membrane surface. Their study showed that the phenomenon of adsorption to the membrane surface is attributable to the strong force between the adsorbate and the CTA membrane surface. They suggested that the adsorbed layer should be considered as a mixed structure, which consists of the region on the membrane and inside the mesh, instead of as monolayer coverage. Figure 8.13 shows a model of a polyelectrolyte-adsorbed form on a CTA membrane.

Bowen et al. [48] described the use of AFM to study the deposition of bovine serum albumin onto the surface of two polymeric membranes of similar molecular weight cutoffs but different surface roughness and surface chemistry (ES404 and EM006 PCI membranes, UK). BSA-coated probes were used to quantify the interaction of BSA with both clean and BSA-fouled membranes. They observed that for the hydrophilic membrane, EM006, the surface roughness was found to increase less with increasing BSA concentration than the hydrophobic, ES404, membrane. The increase in surface roughness was correlated with the flux loss. By using a silica colloid probe, Hilal and Bowen [49] studied the rejection of colloids by membrane pores. Their study provided experimental evidence of the influence of electrostatic double layer interactions on rejection at the pores of a microfiltration membrane.

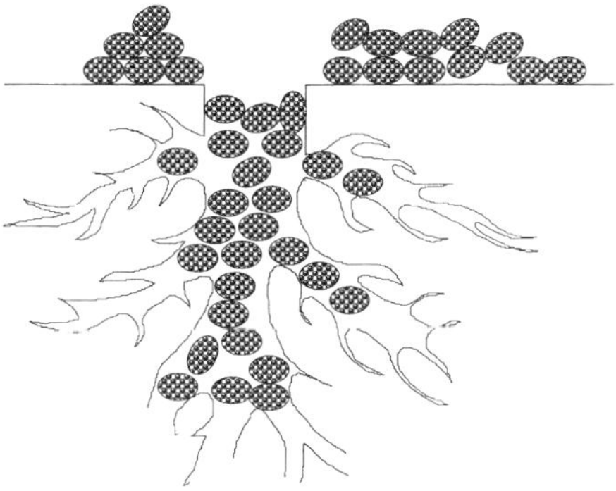

Fig. 8.13. Model of polyelectrolyte-adsorbed form on CTA membrane. Reprinted from [47]. Copyright 1997, with kind permission from Elsevier

8.4 AFM Study of the Dry and Wet Surfaces of the Membrane

We have attempted to indicate, as much as possible, if the membranes studied by AFM were in a dry or wet state, since membrane morphology changes considerably from dry to wet. The following experimental data give evidence for such changes.

A dense homogeneous cellulose acetate membrane was studied by AFM in its dry and wet state [50]. Figures 8.14, 8.15, and 8.16 show the three-dimensional AFM images of the CA membrane in a dry state, after contacting the surface with a few drops of water, and after drying the water-treated membrane at room temperature for 4 days, respectively. Table 8.10 summarizes the mean nodule aggregate sizes and the roughness parameters of the surfaces.

On comparing Figs. 8.15 and 8.16, it seems that the nodule aggregates have swollen by wetting and are fused with each other, forming larger nodular aggregates. This is reflected by the increases observed in the nodule size and the roughness parameter. Interestingly, there is only little change, both in the figures and in the data shown in Table 8.10, when the water-wet surface is dried at room temperature. In other words, the swelling of the nodule aggregates was irreversible within the reported experimental conditions. Most likely, the fusion of the nodule aggregates is responsible for the high salt rejection in RO as well as high selectivity in gas separation of the cellulose acetate membrane. A thorough discussion of RO transport in relation to membrane surface morphology was also made by the same group [50] and by Ward et al. [51].

Table 8.10. Roughness parameters and nodular aggregate of dry and water-treated CA membranes

CA membrane	Roughness, R_a (nm)	Nodule aggregate size (nm)		
		Mean	Max.	Min.
Dry	67.12	174.7	235.9	117.9
Water-treated [a]	88.70	272.2	373.0	116.8
Water-treated membrane dried for 4 days at room temperature	84.60	265.0	333.4	126.9

[a] Excess of water from the surface after 2 h was removed by absorption with filter paper

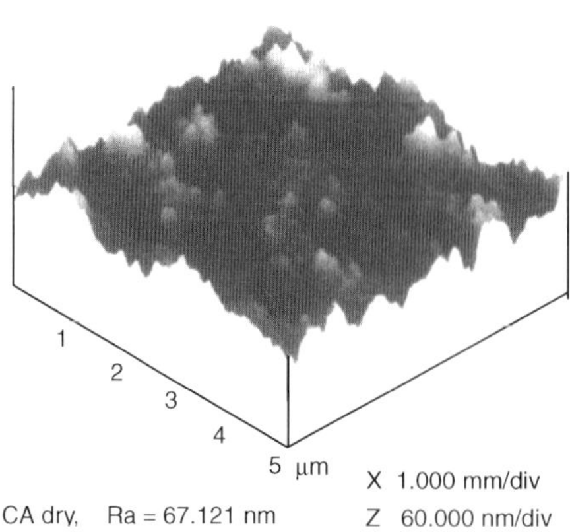

Fig. 8.14. AFM image of dry CA membrane. Reprinted from [50]. Copyright 2004, with kind permission from Elsevier

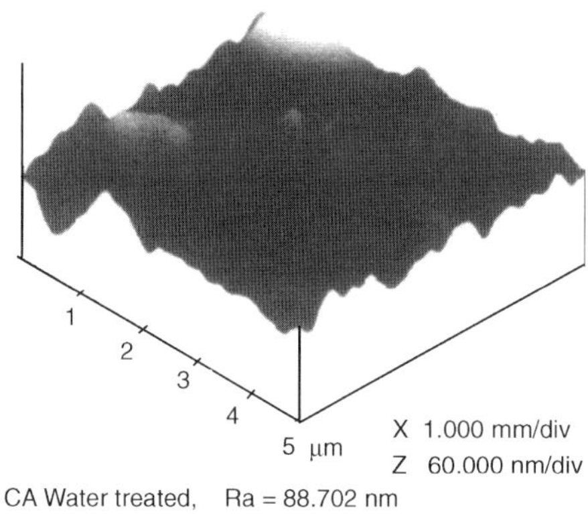

Fig. 8.15. AFM image of CA membrane treated with water for 2 h. Reprinted from [50]. Copyright 2004, with kind permission from Elsevier

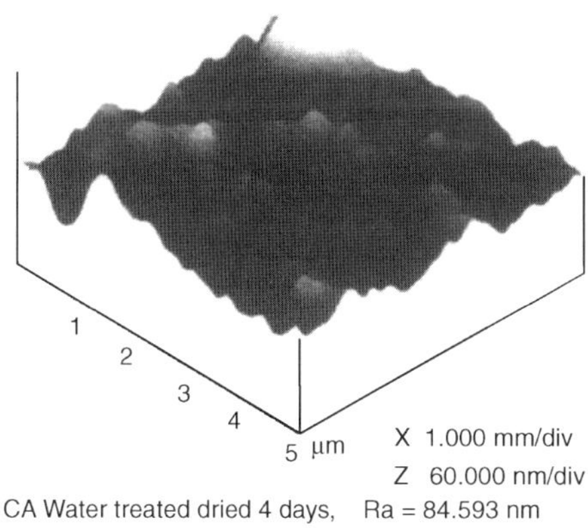

Fig. 8.16. AFM image of CA membrane treated with water for 2 h and dried at room temperature for 4 days. Reprinted from [50]. Copyright 2004, with kind permission from Elsevier

8.5 Summary

Some of the important conclusions drawn from the survey given in this chapter are:

1. The statement of Hirose et al. [1] that higher surface roughness of the polyamide TFC membrane for RO will enhance the membrane flux has been supported by some researchers through their independent experiments. On the other hand, contradictory results have been obtained by others. This may be due to the fact that surface roughness alone cannot be varied independently from other parameters that may also affect membrane flux. One such parameter is obviously the chemistry of the membrane surface.
2. Khulbe et al. [20] and Tan et al. [21, 22] both stated that smoother membrane surfaces, formed as a result of the merging of nodules, would enhance the selectivity of membranes when they were used for gas separation. This concept has not been proven by other researchers since few people have applied the AFM technique to study gas separation membranes. Combined with plasma etching, the AFM technique will provide detailed information on the surface morphology of gas separation membranes, which will definitely influence the future of gas separation membrane design.

3. As already mentioned, the effects of pore size and pore size distribution on selectivity and flux of NF/UF/MF membranes are obvious, which has been observed by many researchers.
4. Higher roughness parameters seem to enhance permeation rates of some pervaporation membranes.
5. A consensus has been reached that a smoother membrane tends to be less susceptible to fouling by particulate matter. Some industrial membranes have been designed based on this concept. However, there is at least one study in which membranes of higher surface roughness were found to be superior to smoother membranes in terms of cleaning capacity. Furthermore, it is not clear if smoother surfaces are also less susceptible to the fouling caused by strong organic adsorption.
6. AFM has been proven to be useful when studying the structure of the colloidal fouling layer formed on the membrane surface. Compact packing of the colloidal layer reduces the membrane flux.

References

1. Hirose M, Ito H, Kamiyama Y (1996) J Membr Sci 121:209
2. Lu X, Bian X, Shi L (2002) J Membr Sci 210:3
3. Wu HR, Wang BF, Guo SH (1983) J East China Univ Sci Technol 2:205
4. Wu J, Wang BF, Wu HR (1992) J East China Univ Sci Technol 18:645
5. Hamza A, Chowdhury G, Matsuura T, Sourirajan S (1997) J Membr Sci 129:55
6. Stamatialis DF, Dias CR, Norberta de Pinho M (1999) J Membr Sci 160:235
7. Kwak SY, Ihm DW (1999) J Membr Sci 158:143
8. Kwak SY, Yeom MO, Roh IJ, Kim DY, Kim JJ (1997) J Membr Sci 132:183
9. Mohammad AW, Hilal N, Seman MNA (2003) Desalination 158:73
10. Bowen WR, Mohammad AW (1998) Desalination 117:257
11. Chung TS, Qin JJ, Huan A, Toh KC (1992) J Membr Sci 196:251
12. Espinoza-Gömez H, Lin SW (2001) Polym Bull 47:297
13. Khayet M, Feng CY, Matsuura T (2003) J Membr Sci 213:159
14. Feng CY, Khulbe KC, Chowdhury G, Matsuura T, Sapkal VC (2001) J Membr Sci 189:193
15. Khulbe KC, Feng CY, Hamad F, Matsuura T, Khayet M (2004) J Membr Sci 245:191
16. Zhang Y, Shao H, Hu X (2003) J Appl Polym Sci 86:3389
17. Hilal N, Kochkodan V, Al-Khatib L, Busca G (2002) Surf Interface Anal 33:672
18. Teng MY, Lee KR, Liaw DJ, Lin YS, Lai JY (2000) Eur Polym J 36:663
19. Khulbe KC, Matsuura T, Lamarche G, Kim HJ (1997) J Membr Sci 135:211
20. Khulbe KC, Matsuura T (2000) J Membr Sci 171:273
21. Tan JMA, Matsuura T (1999) J Membr Sci 160:7
22. Tan JMA (1999) Ph.D. thesis, University of Ottawa
23. Kapantaidakis GC, Koops GH (2002) J Membr Sci 204:153
24. Kapantaidakis GC, Koops GH, Wessling M (2002) Desalination 145:353
25. Khulbe KC, Feng C, Matsuura T, Kapantaidakis GC, Wessling M, Koops GH (2003) J Membr Sci 226:63
26. Reid BD, Ebron VHM, Musselman IH, Ferraris JP, Balkus KJB Jr (2002) J Membr Sci 195:181
27. Gould SAC, Schiraldi DA, Occelli ML (1997) J Appl Polym Sci 65:1273
28. James PJ, Elliot JA, McMaster TJ, Newton JM, Elliot AMS, Hanna S, Miles MJ (2000) J Mater Sci 35:5111
29. Elliot JA, Hanna S (1999) J Appl Crystallogr 32:1069
30. Hsu WY, Gierke TD (1983) J Membr Sci 13:307
31. Soresi B, Quartarone E, Mustarelli P, Magistris A, Chiodelli G (2004) Solid State Ionics 166:383
32. Wang F, Hickner M, Kim YS, Zawodzinski TA, Thomas A, McGrath JE (2002) J Membr Sci 197:231

33. Borges JP, Godinho MH, Martins AF, Stamatialis DF, De Pinho MN, Belgacem MN (2004) Polym Compos 25:102
34. Barzin J, Feng C, Khulbe KC, Matsuura T, Madaeni SS, Mirzadeh H (2004) J Membr Sci 237:77
35. Cohen RD, Probstein RF (1986) J Colloid Interface Sci 114:194
36. Elimelech M, Zhu X, Childress AE, Hong S (1997) J Membr Sci 127:101
37. Freger V, Gilron J, Belfer S (2002) J Membr Sci 209:283
38. Vrijenhoek EM, Hong S, Elimelech M (2001) J Membr Sci 188:115
39. Zhang W, He G, Gao P, Chen G (2003) Sep Purif Technol 30:27
40. Bowen WR, Doneva TA, Yin H (2002) Desalination 145:39
41. Dobrynin AV, Colby RH, Rubinstein M (1995) Macromolecules 28:1895
42. Bowen WR, Doneva TA, Yin HB (2002) J Membr Sci 206:417
43. Bowen WR, Doneva TA, Yin HB (2001) J Membr Sci 181:253
44. Riedl K, Girard B, Lencki RW (1998) J Membr Sci 139:155
45. Huisman IH, Prádanos P, Hernández A (2000) J Membr Sci 179:79
46. James BJ, Jing Y, Chen XDJ (2003) J Food Eng 60:431
47. Murata T, Tanioka A (1997) J Colloid Interface Sci 192:26
48. Bowen WR, Doneva TA, Austin J, Stoton AG (2002) Colloids Surf B 27:103
49. Hilal N, Bowen WR (2002) Desalination 150:289
50. Khulbe KC, Hamad F, Feng C, Matsuura T, Khayet M (2004) Desalination 161:259
51. Ward RA, Feldhoff PW, Klein E (1985) Membrane materials for therapeutic application in medicines. In: Lloyd DR (ed) Materials science of synthetic membranes. ACS Symposium Series 269. American Chemical Society, Washington, DC, p 99

Unfortunately, there is no such membrane which can separate happiness and sorrow of our life. *T. Matsuura*

Subject Index

Printing: Krips bv, Meppel, The Netherlands
Binding: Stürtz, Würzburg, Germany